LESSONS IN ROTARY DRILLING
Unit V — Lesson 2

Spread Mooring Systems

Published By

PETROLEUM EXTENSION SERVICE

The University of Texas at Austin

in cooperation with

INTERNATIONAL ASSOCIATION
OF DRILLING CONTRACTORS
Houston, Texas
1976

©1976 by The University of Texas at Austin
All rights reserved
Published 1976. Second Impression 1978
Printed in the United States of America

TABLE OF CONTENTS

FOREWORD

Chapter I. INTRODUCTION
- Historical Background ... 1
- The Offshore Environment ... 6
- Mooring Patterns ... 8
- Basic Principles of Mooring Systems ... 9

Chapter II. MOORING SYSTEM COMPONENTS
- Overview ... 13
- Anchors ... 13
- Chain ... 16
- Wire Rope ... 20
- Connecting Elements ... 24
- Winches ... 27
- Windlasses ... 28
- Mooring Line Tension Measuring Devices ... 29
- Pendant Lines and Mooring Buoys ... 32
- Anchor Handling Boats ... 33

Chapter III. PLACING AND RECOVERING MOORINGS
- Planning and Organization ... 35
- Moving Onto Location ... 35
- Running Out An Anchor ... 37
- Setting An Anchor ... 37
- Preloading and Pretensioning Anchor Lines ... 38
- Piggybacking Anchors ... 38
- Mooring System Adjustments in Rough Weather ... 38
- Pulling An Anchor ... 39
- Moving Off Location ... 39

GLOSSARY ... 43

REFERENCES ... 45

FOREWORD

The Committee on Education and Training of the International Association of Drilling Contractors is the sponsor of five units of elementary training material on drilling oil and gas wells, entitled *Lessons in Rotary Drilling*. The entire series represents two years of home study, although many will find it possible to complete the material in a fraction of that time.

Lesson 2 of Unit V, the Offshore Technology unit, is entitled *Spread Mooring Systems*. Other lessons in Unit V are listed on the inside front cover. The lessons in Units I, II, and III are listed on the inside back cover.

The purpose of *Spread Mooring Systems* is to orient the man who is new to working on a floating rig. The specific problems of holding on location and the equipment involved to maintain position above a well are described and discussed. It is assumed that men who are new to offshore operations are already familiar with drilling procedures used on land.

                                                   W. E. Boyd
                                                   Petroleum Extension Service
                                                   University of Texas at Austin

Austin, Texas
June 1976

CHAPTER I
INTRODUCTION

HISTORICAL BACKGROUND

The problem of holding a floating vessel near a fixed location on the sea surface is as old as sea-faring civilization. Basic principles and concepts have changed little since the most ancient times; however, technologies and applications have improved markedly.

The earliest mooring systems consisted of natural fiber ropes attached to anchor stones. Various types of anchor stones dating back as far as 1600 B.C. have been recovered from Egyptian tombs and the Mediterranean Sea floor. Metal anchors were introduced by the year 800 B. C. when bronze anchors were cast on the island of Malta. By 300 B. C. iron anchors were common to ships of the Athenian navy. As shown in figure 1, some of these early anchors contain the features of a relatively modern anchor, the Admiralty anchor. Rapid development of anchors in the years following the Industrial Revolution culminated in the "stockless" anchor of the early twentieth century.

Today, interest in anchor development has been revived by the floating drilling industry, whose offshore operations require anchors with improved holding power in all types of bottom conditions, including both mud and sand.

As anchors improved over the years, so did anchor and mooring lines. Natural fiber rope is seldom used in modern mooring applications in view of its light weight, low breaking strength, and susceptibility to biological attack. With light weight and low breaking strength, fiber rope is prone to jerk taut and part in severe weather conditions. By comparison, metal chain has much greater weight and a higher breaking strength. The advantage of pure dead weight is that a chain mooring line forms a catenary curve, a "spring" which will absorb the shock of surge loadings and bring a vessel under control without damage. The earliest known users of chain were the Chinese, who described moorings of iron chain in 2200 B. C. History does not record western use of chain until 950 B. C. when Hiram of Tyre furnished chains, probably of brass, for the ships of King Solomon. Chain has the obvious military advantage that it cannot be cut easily; however, the wrought iron chain of early times was so heavy that its weight displacement could often be used more effectively for additional armaments or cargo. Widespread use of chain became possible in the nineteenth century when chain making technologies and designs improved significantly. In 1812, studs were forged in chain links to stiffen them and prevent them from tangling. The proportions of modern-day chain links and studs have evolved over many years of experimentation.

Another evolutionary product is wire rope, which represents a marriage of modern steel technology with ancient rope-making traditions. In comparison with chain, wire rope has a much greater strength-to-weight ratio, but is much more susceptible to damage by abrasion, corrosion, and general abuse. In mooring applications where technically feasible, chain is usually favored over wire rope. However, wire rope is used extensively in deep-water mooring systems where high strength-to-weight ratios are important. Composite mooring systems, including both wire rope and chain, combine the advantages of both.

When a cargo vessel or naval ship is moored, it normally swings with the seas. Mooring problems are difficult with a floating drilling vessel because it must be held within a few feet of a fixed location for a considerable length of time. The task of keeping a vessel stationary for long periods of time was first tested by the early drilling tenders which appeared in the Gulf of Mexico following World War II. As shown in figure 2, the tender

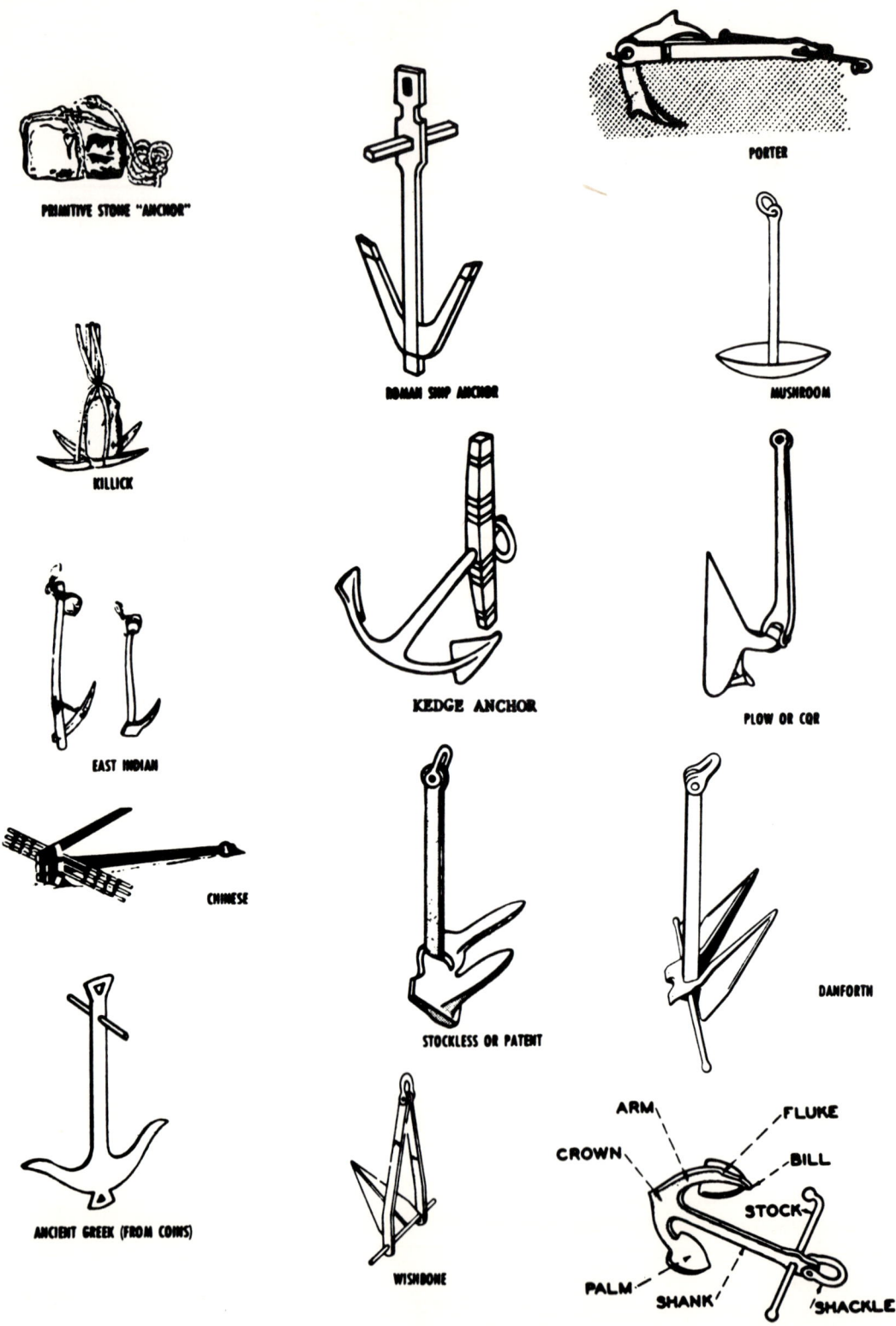

Figure 1. Anchors of Historical Interest

was moored adjacent to a small platform which held the derrick, drawworks, and engines. Crew quarters, supplies, and remaining drilling equipment were contained on the tender, with transfers being achieved by means of a bow ramp joining the tender and the platform. In rough weather, the bow ramp would be disconnected and the tender pulled back from the platform.

In general, the drilling tenders were modified U.S. Navy LST's with all propulsion equipment removed. Several mooring systems were developed and used for the tenders, including anchors and pile clusters in various combinations. One mooring system is shown in figure 3, an eight-line spread mooring pattern.

Use of drilling tenders offshore declined following the development of completely mobile drilling units and movement of offshore oil exploration into deeper, rougher waters where platforms and operations were much more costly. The first mobile drilling units were submersible barges, originally used in the coastal inland waters of Louisiana and California. As early as the 1930's, wells were drilled at inland marine locations using barge-type rigs which could be transported from one location to another via dredged canals. Submersibles carried drilling equipment on a deck which was raised above a lower flotation hull by posts or columns. Once the submersible was moved onto location, the lower flotation hull would be flooded and the rig would settle on bottom. Submersible drilling units for offshore use suffered obvious water depth limitations, the largest ever built being restricted to 175 feet of water.

To overcome water depth limitations, spread mooring techniques used by drilling tenders were adapted, creating the semi-submersible unit capable of drilling in a partially submerged condition. Among the earliest semi-submersible drilling units were the Socal Western Explorer (1955) and the

Figure 2. Drilling Tender at Work

To drill in deep waters, drill ships sometimes employ "dynamic positioning" systems, enabling the ship to renew its position relative to the well bore and make corrections by activating propulsion thrusters. The total thrust which the system can muster is fixed, and any environmental force exceeding this amount will blow the ship off location. Instrumentation for dynamic positioning systems is complex and costly, as are the operating procedures. Fuel costs alone total thousands of dollars per day. In view of the limited applications of dynamic positioning, the scope of this discussion is restricted to spread mooring systems.

Although there are considerable variations in engineering design, the basic principles and equipment used in spread moorings are the same for both semi-submersibles and drill ships.

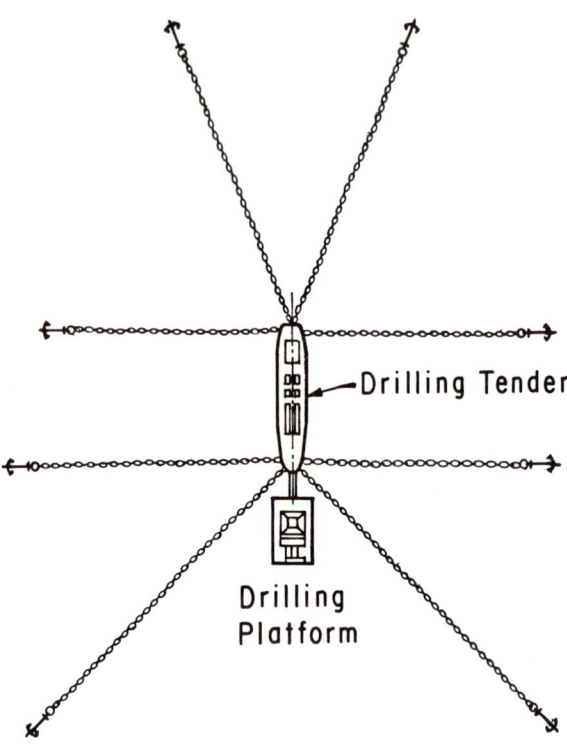

Figure 3. Eight-Line Spread Mooring System

Bluewater I (1961). Figure 4 is a basic outline of a semi-submersible and the loads and environmental forces which are exerted on it.

Semi-submersibles are bulky and historically have been handicapped by slow towing speeds. Therefore, long rig moves are often time consuming and costly. This problem has been alleviated to some extent by including self-propulsion capability in the newer semi-submersibles.

Where a high degree of mobility is desired, floating drilling operations may be conducted from a drill ship, which is self-propelled and can move at a better speed than semi-submersibles, but is less transparent to waves and seas. Thus, it is subject to more vessel motion and more down time due to hostile weather. A typical drill ship configuration is shown in figure 5; note the mooring lines and arrangement of the underwater drilling equipment.

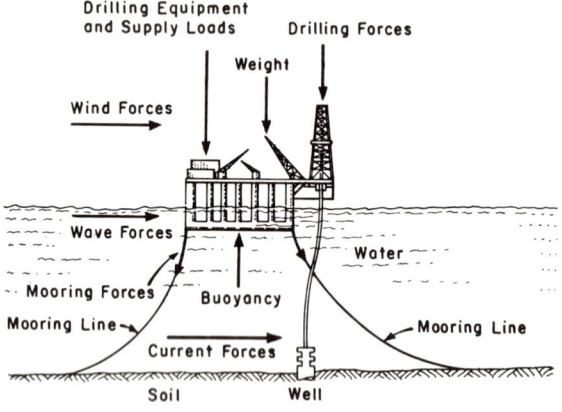

Figure 4. Loads and Environmental Forces on a Semi-submersible Drilling Rig

4

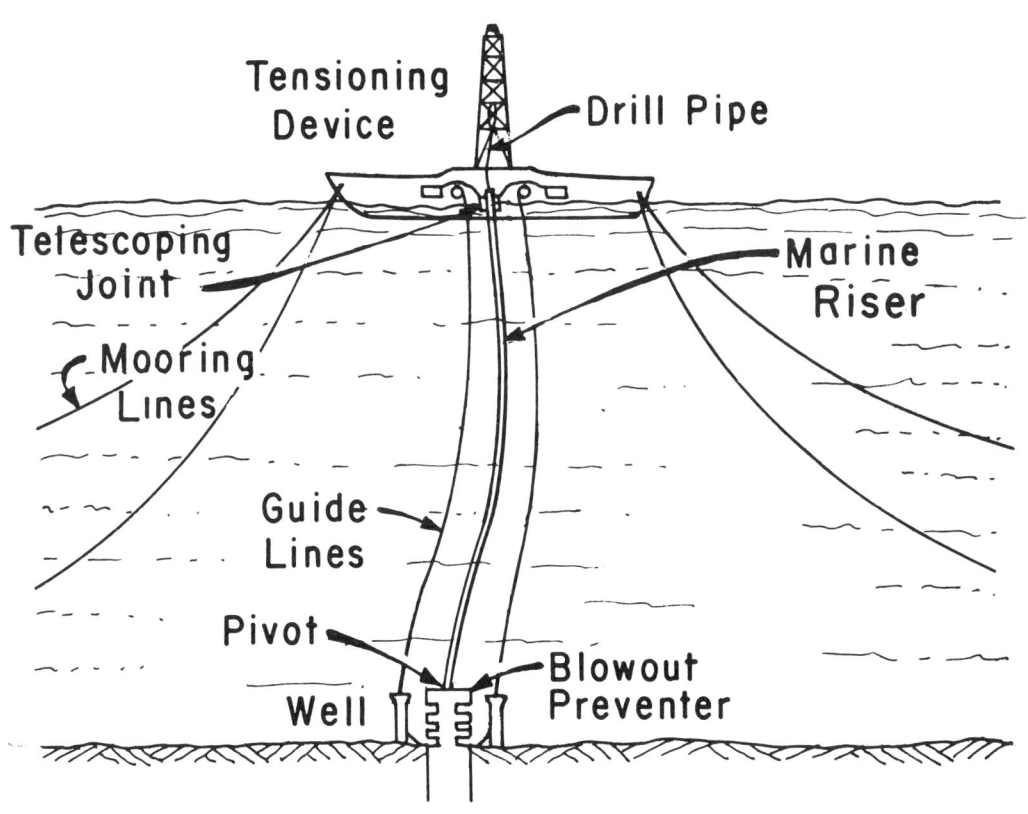

Figure 5. Drill Ship Arrangement of Mooring Lines and Drilling Assembly

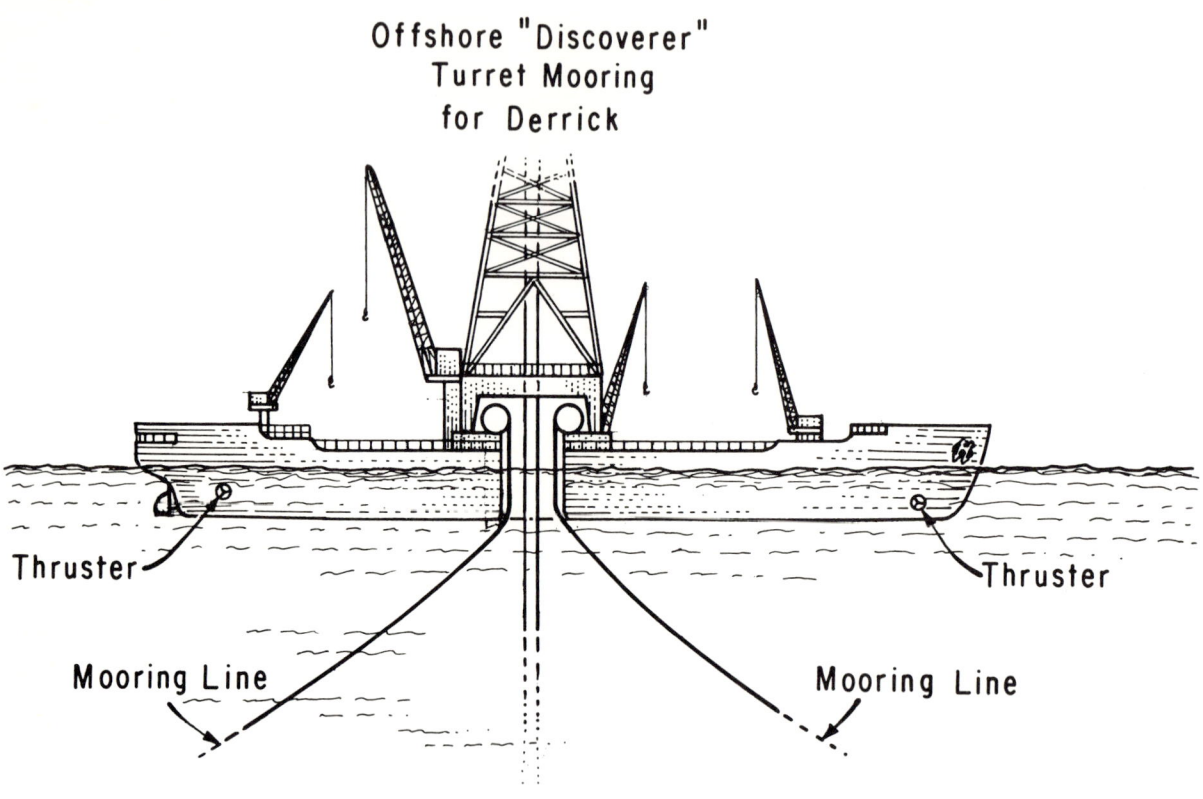

Figure 6. Turret-type Mooring Arrangement

Figure 6 illustrates a unique mooring design, turret mooring, which is used by The Offshore Company "Discoverer" class vessels. The mooring lines are wire ropes which are spooled onto winches mounted on the turret, which is located at the center of the vessel. Since all mooring lines are connected to the turret, the vessel is free to rotate about the turret-wellhead axis and head into oncoming seas, regardless of direction. Thrusters are provided at the bow and the stern to assist in maintaining a desired heading.

THE OFFSHORE ENVIRONMENT

Drilling an exploration well offshore requires that a floating drilling rig remain on location from 30 to 120 days. During that time, the mooring system must prevent excessive movement of the vessel so that drilling operations can proceed as continuously as possible. Vertical movements of the vessel are absorbed by heave compensators and/or slip joints in the marine riser. Horizontal offset movements from the centerline of the well are permitted by a ball joint at the top of the blowout preventer (BOP) stack. However, excessive offsets may endanger the marine riser, the drill pipe, and the BOP assembly. As shown in figure 7, drilling operations cannot continue when the ball-joint angle exceeds a certain value or damage will occur to the drill pipe and the BOP assembly. To make hole in normal drilling operations, the ball-joint angle should be less than 4°, corresponding to a horizontal vessel offset from the well which is about 3% of water depth. Maximum operating conditions are a 10° ball-joint angle and 6% vessel offset. Under maximum operating conditions, drilling has been discontinued; however, the rig may continue circulating

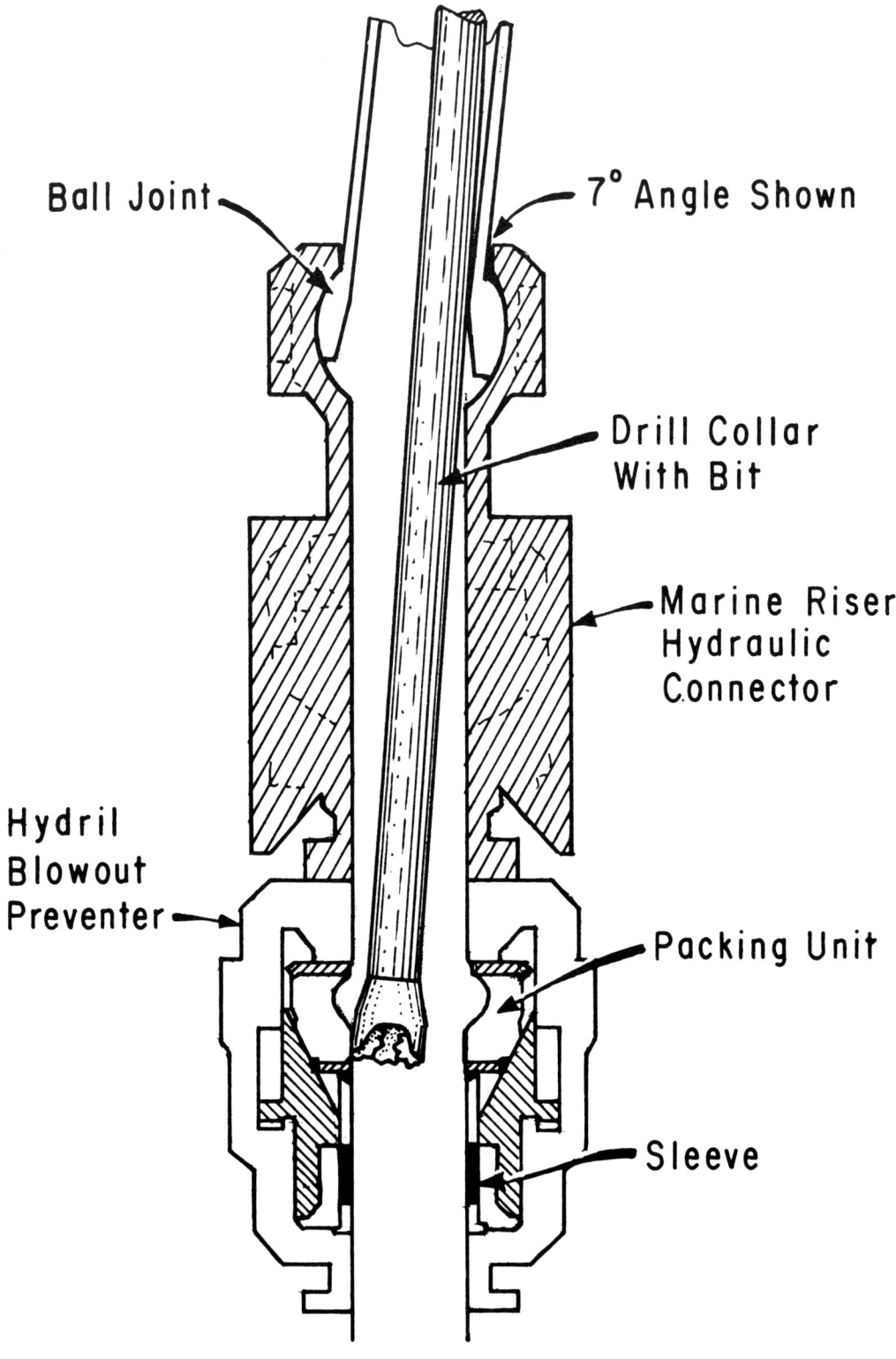

Figure 7. Wellhead Flexible Connection, Connector, and Blowout Preventer Arrangement

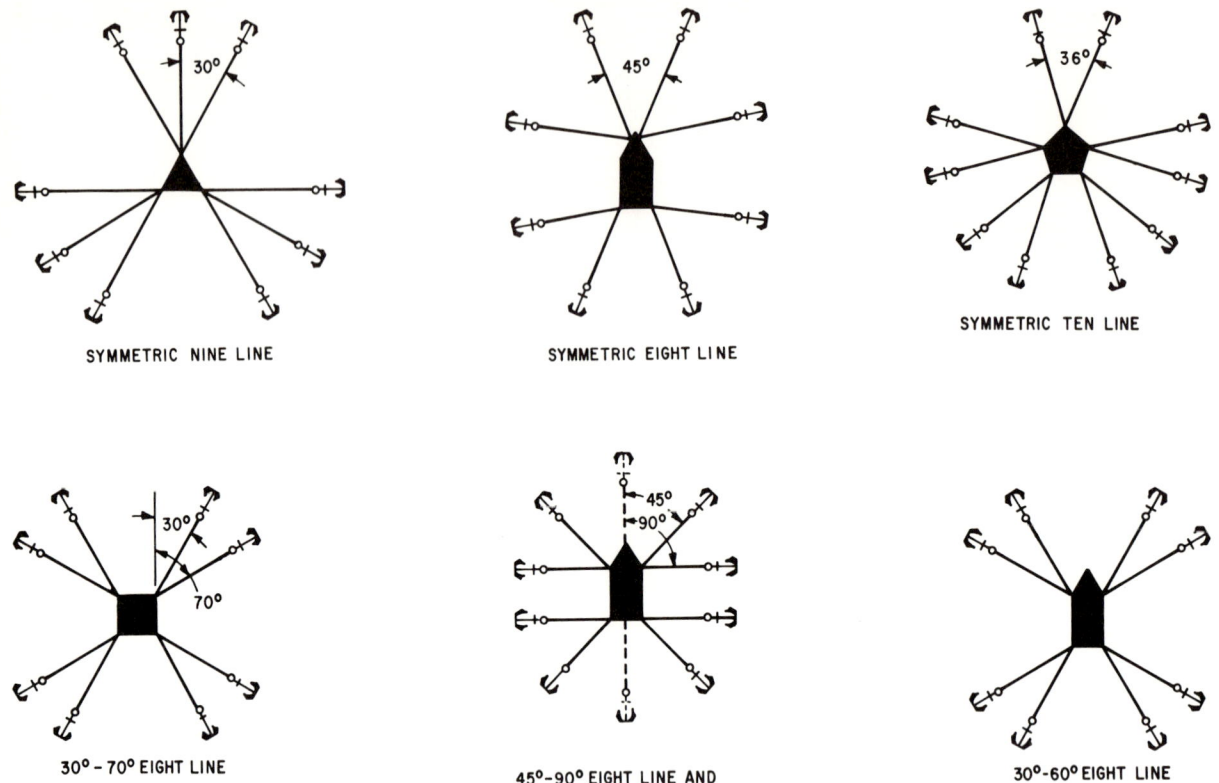

Figure 8. Typical Spread Mooring Patterns

mud and pulling pipe while waiting on weather changes. Beyond maximum operating conditions, sea water must be used to displace heavy drilling mud from the marine riser in order to prevent the riser from being overstressed and lost. The marine riser can remain connected to the BOP assembly up to about a 10% vessel offset, at which point it must be disconnected and the rig prepared to ride out the rough weather.

A floating drilling rig and mooring system which will merely survive hostile environments is insufficient. Rigs have been designed to withstand survival conditions of more than 110-foot waves simultaneously with a three-knot current and 100-knot wind, which is equivalent to a wind speed of 116 miles per hour. The rig and mooring system must also resist movement in moderately severe conditions in order to continue normal drilling operations and economically complete an exploration well. Specified criteria call for normal drilling operations to continue in 15 to 30-foot waves, 30 to 50-knot winds and a one to two-knot current.

MOORING PATTERNS

In the open sea, weather can attack a vessel from any direction, although certain directions may be preferred in some locations. Therefore, it is necessary to array mooring lines radially away from the vessel in several directions. The number of lines and the pattern chosen will depend on the exposure of the vessel to environmental forces, preferred weather directions, maximum allowable forces in the mooring lines, on board anchor handling equipment, and other considerations. Figure 8 shows mooring patterns which have been used for various classes of drilling vessels. The symmetric nine-line system, for example, is used by the SEDCO "135" and ODECO "Ocean Driller" class vessels.

BASIC PRINCIPLES OF MOORING SYSTEMS

Environmental forces on the drilling vessel are counterbalanced by restoring forces which are supplied by tensions in the mooring lines. In turn, the mooring line tensions are transmitted to anchors on the ocean bottom. In order to gain a physical understanding of spread mooring systems, it is necessary to consider the basic principles which govern the behavior of anchors and mooring lines.

Anchor Holding Power

Figure 9 gives the nomenclature of a basic type of anchor. Figure 10 shows different observed types of anchor behavior. The top illustration in figure 10 depicts an anchor when performing properly. The shank is horizontal, the flukes are tripped to a preset angle, the anchor has buried itself, and increased horizontal pull will cause the anchor to dig deeper. The mooring line should also be horizontal, although some angle from horizontal is acceptable, up to a maximum of 6°. Any significant vertical pull on the anchor beyond this amount will cause it to pull out. In soft mud bottom, the flukes may fail to trip or the anchor may pull out with a large ball of mud covering the fluke and crown assembly. This failure to hold is called "balling up" (figure 10).

Holding power of an anchor is often expressed in terms of a "holding power ratio," which is defined as the mooring line tension on the anchor divided by the weight of the anchor in the air. An efficient anchor should have a high holding power ratio; that is, it should develop a maximum of holding power for a minimum of weight. Ideally, the holding power ratio for a given anchor should exceed 10 for all types of bottom conditions ranging from hard sand and clay through soft mud. A more conservative rule of thumb is that holding power may be estimated as three times anchor weight.

R. W. Beck (1972) confirmed that maximum holding power is sensitive to fluke angle setting. For a soft bottom, the fluke angle should be set at about 50°. Where the bottom is hard, the fluke angle should be closed down to approximately 30°. M. A. Childers (1972) notes that holding power has been improved in hard bottom areas by sharpening the fluke edges with a cutting torch and increasing the length of the stock. This prevents the anchor from turning over and gives the flukes better digging power.

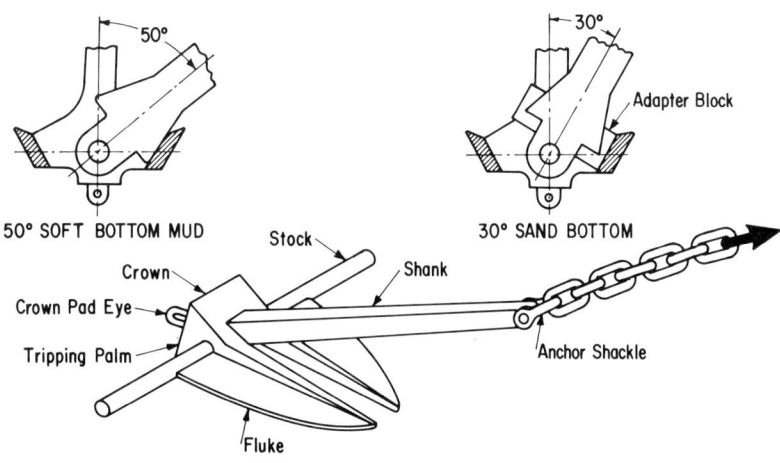

Figure 9. Basic Type Anchor with Nomenclature of Parts

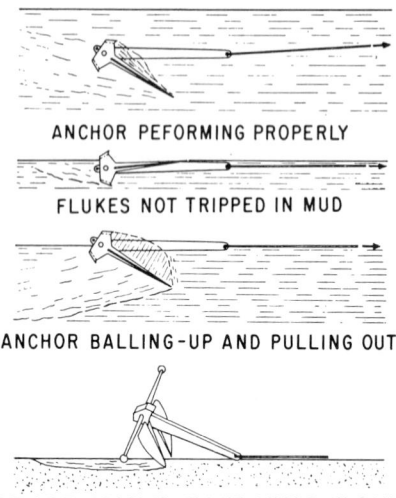

Figure 10. Anchor Behavior Depends on Bottom Conditions

Catenary Curves

Under the force of its own weight, the suspended length of a mooring line will fall into a shape known as a catenary curve. Figure 11 shows a catenary curve for a semi-submersible. Note that extra line is provided at the bottom to make sure that all pull on the anchor is horizontal and that it will not be pulled out vertically. The ratio of the total length of the mooring line to the water depth is the "scope." Mooring lines for floating drilling vessels commonly have minimum scopes of 5.0 to 7.0, or total lengths which are five to seven times water depth.

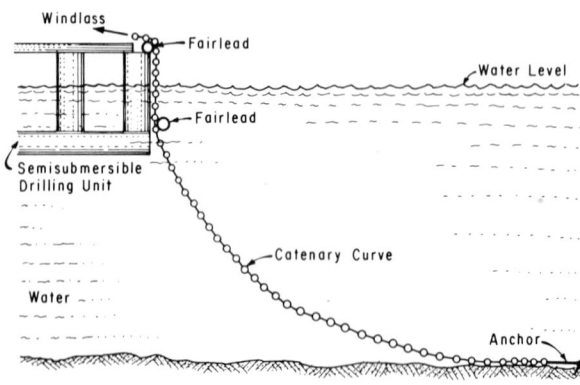

Figure 11. Catenary Curve

In addition to the chain's weight, wave and current forces act on the mooring line. However, the effects are relatively small and are usually neglected for engineering purposes. Mooring-line tension which is calculated by neglecting these forces generally agrees well with tension measured in the field. To calculate mooring-line tension, it is assumed that a mooring line is perfectly flexible, that is, it has no flexural strength. Therefore, no bending can exist along the line, and all loads must be carried in simple tension.

Figure 12 shows the suspended length "S" of the mooring line between the fairlead and the point where it touches bottom. To balance horizontal forces on this length of line, the horizontal force "H" at the fairlead must equal the total line tension at the bottom, inasmuch as the chain is assumed to lie flat as it approaches the anchor. The weight of line between the bottom and the fairlead must be supported by the vertical component "V" of total line tension at the fairlead. Therefore, "V" is given by the expression:

$$V = wS$$

where V = vertical component of mooring line tension at the fairlead, lbs.
    w = submerged weight, lbs./ft.
    S = suspended length of mooring line, ft.

The mathematics of calculating mooring line tension are complex; however, two simple and useful equations emerge. These are:

$$T = H + wd \quad (1)$$

$$S = \sqrt{d\left(\frac{2H}{w} + d\right)} \quad (2)$$

where T = mooring line tension at the fairlead, lbs.
    H = horizontal component of mooring line tension at the fairlead, lbs.
    d = depth from the fairlead to the bottom, ft.

Equation (1) states that the total mooring line tension (T) at the fairlead is equal to the horizontal component of tension (H) plus an additional force which is equal to the submerged weight of a completely slack line in water depth (d). Equation (2) relates the suspended length (S) to the horizontal force H. As H increases, the line will be picked up off bottom and S will increase.

If a vessel is anchored by a single mooring line, the horizontal mooring line force H is equal to the horizontal environmental force imposed by wind, waves and current. If the

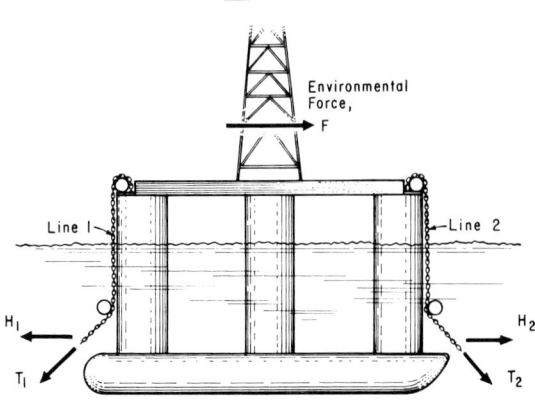

Figure 13. Environmental Force Applied to a Semi-submersible

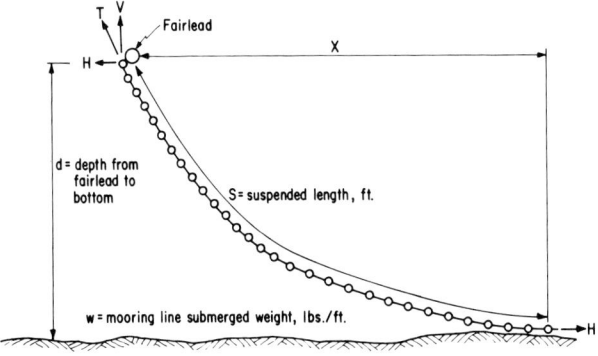

Figure 12. Parameters Required to Calculate Mooring Line Tension

vessel is anchored by multiple lines, the environmental force is shared between them, as shown in Figure 13. In order to maintain an equilibrium, the environmental force F must be balanced by a net restoring force from the mooring lines.

$$F = H_1 - H_2 \qquad (3)$$

where F = total environmental force imposed on vessel by waves, wind and current

H_1 = horizontal component of tension in mooring line 1

H_2 = horizontal component of tension in mooring line 2

The important point to note in Equation (3) is that part of the tension H_1 is taken up in counterbalancing the tension H_2. When F is zero, the mooring line tensions simply balance each other.

In rough weather, the tension in windward — upwind — lines, such as Line 1, can approach maximum allowable load. One means of lowering the tension in Line 1 is to slack Line 2, thereby reducing H_2. Figure 14 shows the improvement in restoring force that can be achieved by slacking two leeward — downwind — lines in an 8-line mooring pattern for a vessel of the ODECO "Ocean Voyager" class.

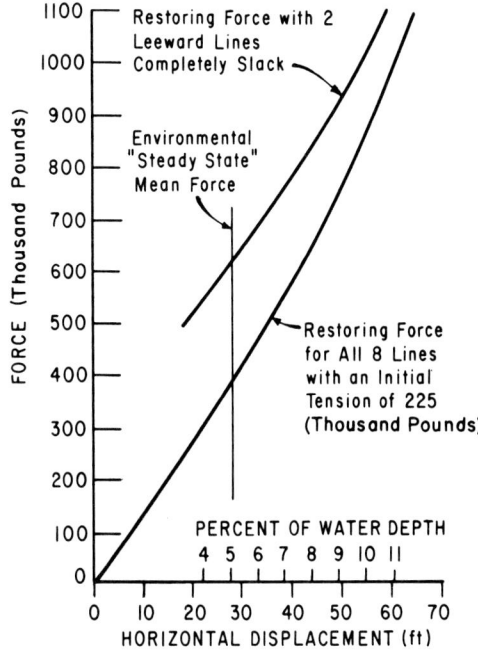

Figure 14. Restoring Forces for a Semi-submersible Due to Horizontal Displacement from the Well

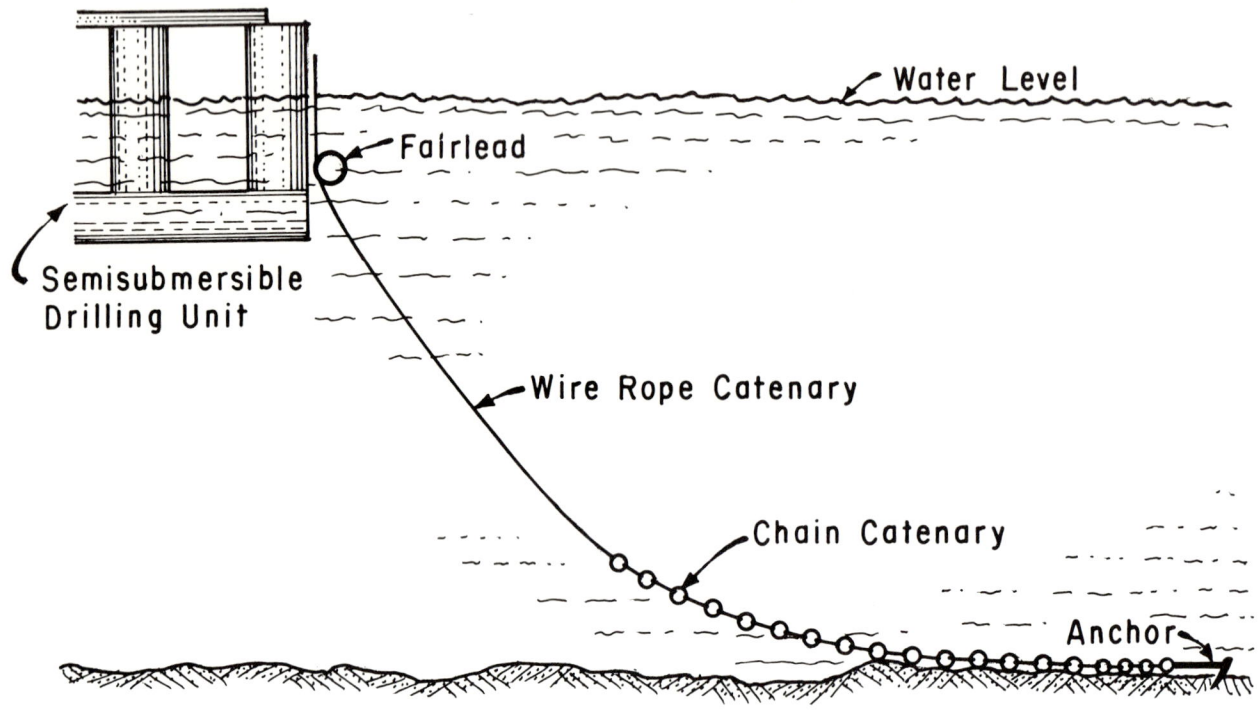

Figure 15. Combination Wire Rope and Chain Mooring Line

A further result of Equation (1) is that the weight and breaking strength of a mooring line limit the water depth in which it can be used. The maximum horizontal restoring force H_{max} will be the difference between the maximum allowable tension in the line T_{max} and the slack line weight wd.

$$H_{max} = T_{max} - wd \qquad (4)$$

Therefore H_{max} decreases as d increases (provided T_{max} and w are fixed) for a given type of mooring line. Decreased H_{max} means that less environmental force can be resisted by the mooring system without excessive stresses.

Equation (4) typifies the limits of deepwater use of chain, which has a very low strength-to-weight ratio. Therefore, many deepwater moorings are designed with lighter, stronger wire rope. However, the disadvantage of lightness is that the catenary curve, length S, may be too long to be economical. Solving this dilemma requires the use of composite mooring lines such as those shown in figure 15. Light, strong line (wire rope) at the top and heavy line (chain) at the bottom assures mooring line tensions are transmitted horizontally to the anchor. A practical difficulty with composite systems is that special equipment and procedures must be employed when using two types of line simultaneously.

CHAPTER II
MOORING SYSTEM COMPONENTS

OVERVIEW

Figures 16 and 17 show two mooring line arrangements which are in practical use. One is a chain configuration which is commonly used on semi-submersible vessels, and the other is a composite wire rope and chain configuration which is used on The Offshore Company "Discoverer" class vessels. Both arrangements contain practically all components which will be found in spread mooring systems: anchors, wire rope, chain, end fittings, handling equipment such as winches and windlasses, and buoys. Not shown are anchor handling boats which run out and position the anchors.

Selection of wire rope, chain, or a composite of the two for mooring lines depends on several considerations. Among these are expected line tensions, water depth, durability, handling equipment, storage facilities, and economics. In general, chain mooring lines are better overall for water depths less than 600 feet. For water depths greater than 100 feet, composite lines with wire rope on top and chain on bottom are better than all-chain mooring lines.

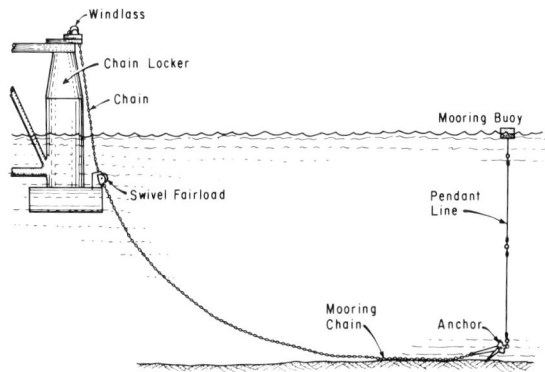

Figure 16. Anchor Chain Mooring Line and Related Equipment

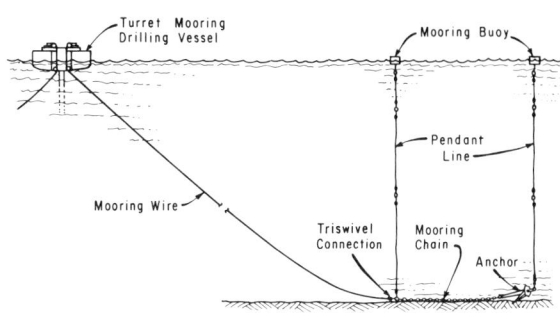

Figure 17. Combination Wire Rope and Anchor Chain Mooring Line

In rough weather areas, such as the North Sea, significant fatigue damage to mooring lines may occur. Fatigue is caused by cyclic variations in mooring line tension about some mean value, and damage from fatigue is greatest when both the mean line tension and the cyclic variations are large. Variations in line tension result from vessel motions caused by environmental forces. Inasmuch as semi-submersibles usually exhibit less motion than drill ships, their mooring systems generally suffer less fatigue damage and have longer service lives.

ANCHORS

Several different types of anchors are in use on floating drilling vessels. Figures 18 through 25 illustrate various types which have been used in this service. The U.S. Navy Light Weight (LWT) anchor and the Danforth anchor are traditional ship-type anchors which have been used extensively. The flukes on these anchors trip to a preset angle when the anchor is dragged on bottom, and also allow for the shank to be dropped through the flukes for ease of handling and storage. Both the LWT and Danforth anchors have good holding power in favorable bottom conditions; however, their holding power in soft mud is poor. In order to maximize holding power,

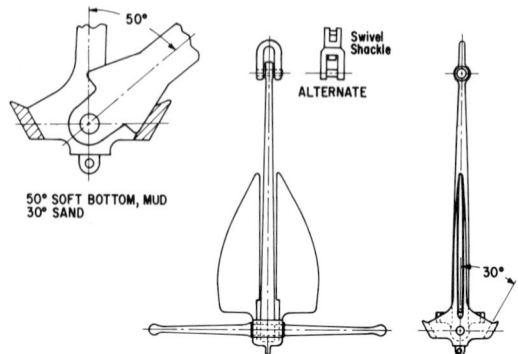

Figure 18. U. S. Navy Light Weight Type Anchor

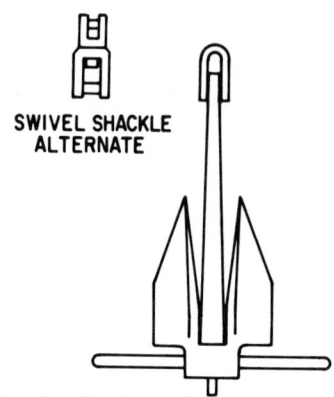

Figure 19. Danforth Anchor, Lightweight Type

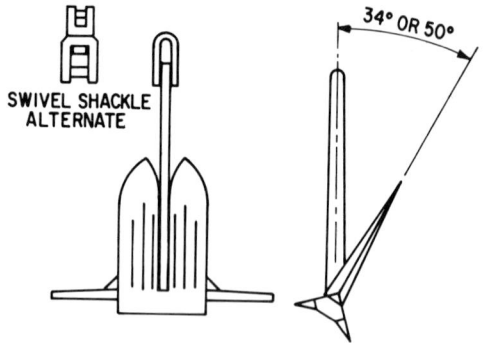

Figure 20. Stato Mooring Anchor

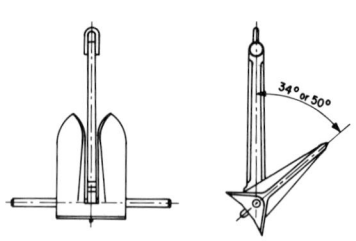

Figure 21. Moorfast Anchor

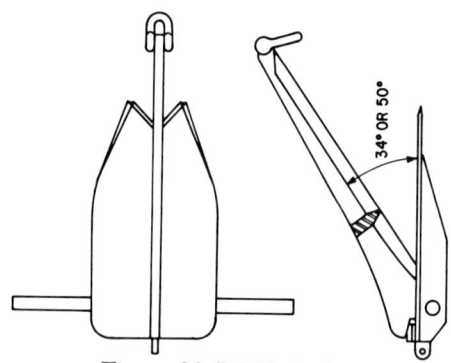

Figure 22. BOSS Anchor

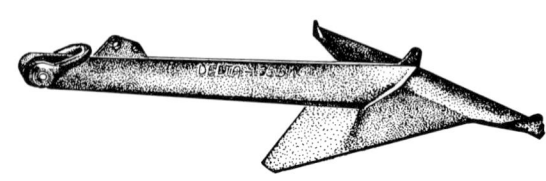

Figure 23. Delta Special Mooring Anchor

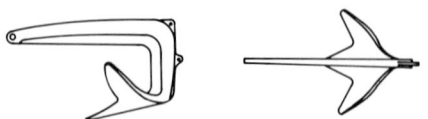

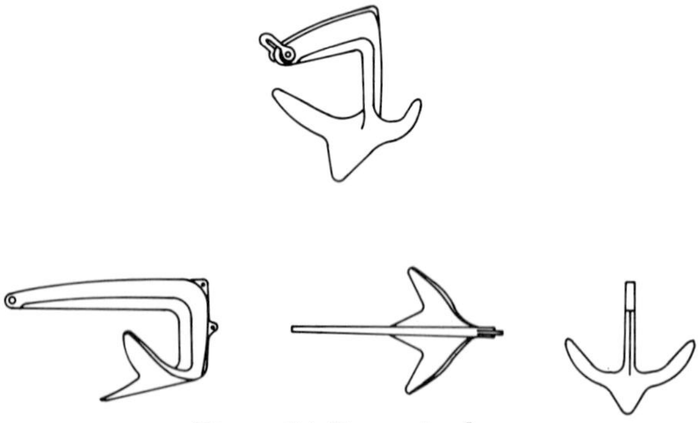

Figure 24. Bruce Anchor

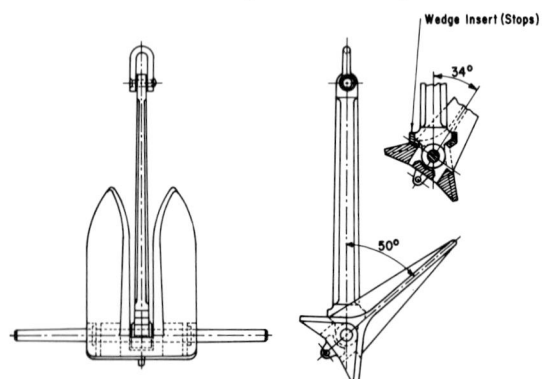

Figure 25. Offdrill II Anchor

14

fluke angle can be adjusted to bottom conditions by means of adapter blocks, sometimes called fluke angle blocks, shown previously in figure 9. Without the adapter blocks, the flukes will trip to an angle of 50° in soft mud bottom; with the adapter blocks, fluke angle can be limited to 30°.

To provide good holding power in both sand and mud bottoms, the U.S. Navy developed the Stato anchor (see figure 20), which has large tripping palms at the rear of the flukes. The ribs which run lengthwise along the broad flukes are not present on the similar Moorfast anchor (figure 21). Tests run by the British Admiralty have indicated that holding power is improved by streamlining and smoothing the flukes. Another commonly used anchor similar to the Stato and Moorfast is the Offdrill II anchor (figure 25).

Many newer anchor designs utilize a single streamlined fluke for maximum fluke area and improved holding power. Usually the angle of the single fluke is fixed to eliminate jamming and tripping failures in mud bottoms. One disadvantage of the single-fluke design, however, is that the shank will not drop through the flukes. Under certain conditions, this can make single-fluke anchors hard to handle on anchor handling boats, and incidents of boat damage have been reported.

To provide excellent holding power and reliability in all types of bottom conditions, Esso Production Research Company developed a single-fluke anchor, the BOSS anchor (figure 22). Fluke angle is adjustable to either 34° for sand bottom or 50° for soft mud bottom. The Delta anchor (figure 23), vaguely resembling the BOSS anchor, has a sharp, thin fluke for maximum digging capacity and minimum soil disturbance. The fluke angle, however, is completely fixed. The Bruce anchor (figure 24) also has a fixed fluke, but a unique configuration. In place of a stock, curved side extensions of the fluke force the fluke to roll upright and dig in when dragged on bottom.

Anchors are normally inspected by anchor handling crews when they are run and retrieved. Usually, anything wrong with an anchor will be obvious; however, all component parts should be examined to make sure that they are in place and holding. Although sometimes difficult to check, the fluke angle blocks should be in place and tight.

Occasionally, hard bottom conditions or vertical pull-out forces restrict the use of normal sea-type anchors. In this situation, it is possible to design and install pile-type anchors, such as the one shown in figure 26. These anchors can be driven or drilled by the rig itself, weather conditions permitting, or can be installed prior to the arrival of the rig. In general, pile-type anchors are expensive and are used only when necessary. Explosively-imbedded anchors may reduce this expense in the future; however, sufficient holding power cannot be obtained at present for floating drilling operations.

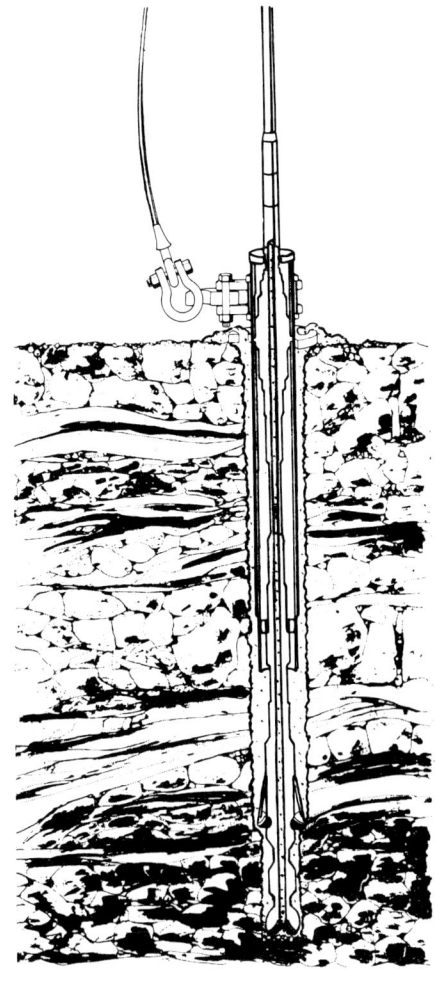

Figure 26. Drilled-Type Anchor Pile

CHAIN

All chain used in mooring floating drilling vessels is of the stud-link type shown in figure 27. The primary purpose of the stud is to prevent the chain from kinking. As an additional benefit, the stud stops distortion of the link under high tension load. Generally, most chain used on floating drilling vessels has strength equivalent to wire rope with a diameter of two to three inches.

Two types of stud-link chain are the "Di-Lok" and the flash-butt welded chain. Di-Lok chain was developed by the U. S. Navy and is manufactured as shown in figure 28. The chain links are forged as separate male and female halves and then are joined together. Flash-butt welded chain is manufactured by heating rolled bars, bending them to the shape of a link, and flash-butt welding the ends together. The studs, which are forged, are then hydraulically pressed into place. For use in drilling operations, the studs should also be welded into the links to prevent subsequent loss. Various grades of flash-butt welded chain are available; however, the one generally used on floating drilling vessels is so-called "Oil Rig" or "Oil Field Stud Link" chain, which has strength and weight equal to Di-Lok. Under load, Di-Lok chain stretches approximately 2.7 times more than flash-butt welded chain. Experimental evidence suggests that this stretching action may result in shorter fatigue life for drilling rig applications.

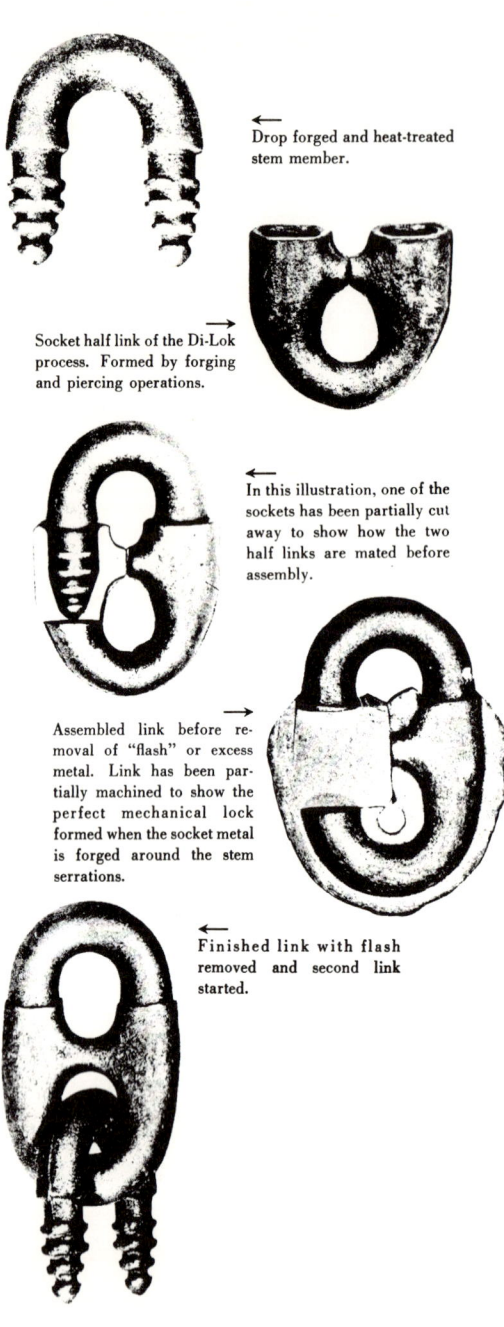

Drop forged and heat-treated stem member.

Socket half link of the Di-Lok process. Formed by forging and piercing operations.

In this illustration, one of the sockets has been partially cut away to show how the two half links are mated before assembly.

Assembled link before removal of "flash" or excess metal. Link has been partially machined to show the perfect mechanical lock formed when the socket metal is forged around the stem serrations.

Finished link with flash removed and second link started.

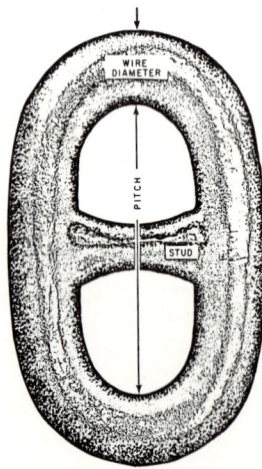

Figure 27. Stud-Link Anchor Chain Nomenclature

Figure 28. Di-Lok Stud-Link Chain Manufacturing Process

16

Design criteria for mooring systems require maximum tension in a chain to not exceed approximately 35% of its breaking strength, or 50% of its proof-load, which are about equal. Proof-load is a load slightly higher than the yield strength of the steel. Breaking strength is the load at which the chain parts. Regular Di-Lok and Oil Rig 3-inch chain have a proof-load of 693,000 lbs. and a breaking strength of 1,045,000 lbs.; therefore, the maximum allowable load for 3-inch chain is 350,000 lbs. Under maximum survival conditions, as much as 50% of the rated breaking strength may be allowed on a limited basis.

Under operating conditions, each line in the mooring system will carry a mean tension which depends on the engineering design. The natural variation in sea conditions will then cause a cyclic variation in load to be superimposed on top of the mean tension. In hostile environments such as the North Sea, where the weather is especially severe, cyclic variations can be large. As a result, chains may fail due to fatigue long before they corrode away or wear out by abrasion. The American Bureau of Shipping uses specifications for removal of chain from service based on ship-type service where wire diameter decreases and chain elongation occurs by processes of corrosion and abrasion. No criteria are available for chain removal when damage is primarily by fatigue. Where fatigue is a problem, usual field practice keeps a chain in service until breakage and general deterioration begin to show. When run and retrieved, the links may show cracks, studs may be loose when struck with a hammer, or studs may be missing completely. Occasionally, the chain may be allowed to break a time or two before being removed from service.

Quality control in the manufacture of chain is extremely important. Standards for fabrication and testing of chain have been provided by the American Petroleum Institute in API Spec. 2F, API Specification for Mooring Chain. The steel used must comply with chemical composition requirements so that it will be of fine grain quality, and must also satisfy tensile strength and ductility test requirements.

Welds in the links should be crack-free and all slag removed. The surface of the chain should be clean and smooth. Chisel marks, marks made by grippers, or hammer marks could serve as stress concentration points for fatigue damage and should be avoided. Studs should have no lugs or protrusions on either end, and should be circumferentially welded into the link on the end opposite the flash weld. Some buyers further specify that the stud be circumferentially welded on both ends to minimize corrosion between the stud and the link. After all welding is completed, the chain is normalized by heating. A combination of time and temperature produces a fine grain structure in the weld and fusion zones of the links. Proper execution of the normalization procedure is extremely important in the life of a chain. Following normalization, specimens of the chain are subjected to a break test, and the entire chain must withstand the specified proof test load. Finally, careful attention should be paid to dimensional tolerances to assure that the chain will fit handling equipment such as windlasses and fairleads properly. Dimensional information for various sizes of Di-Lok and flash-butt welded chain is given in figures 29 and 30.

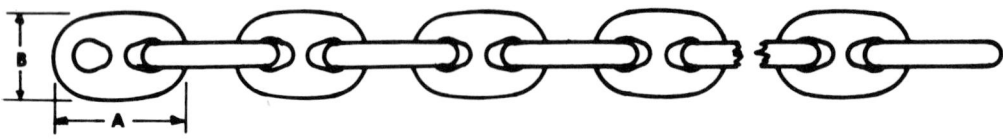

CHAIN SIZE	LINK LENGTH A	LINK WIDTH B	LENGTH OVER SIX LINKS C	LINKS PER 15 FATHOM SHOT	WEIGHT PER 15 FATHOM SHOT APPROX.	DI-LOK PROOF TEST POUNDS	DI-LOK BREAK TEST POUNDS
2	12	7 3/16	52	133	3,525	322,000	488,000
2 1/16	12 3/8	7 7/16	53 5/8	129	3,750	342,000	518,000
2 1/8	12 3/4	7 5/8	55 1/4	125	3,975	362,000	548,000
2 3/16	13 1/8	7 7/8	56 7/8	123	4,215	382,500	579,100
2 1/4	13 1/2	8 1/8	58 1/2	119	4,460	403,000	610,000
2 5/16	13 7/8	8 5/16	60 1/8	117	4,710	425,000	642,500
2 3/8	14 1/4	8 9/16	61 3/4	113	4,960	447,000	675,000
2 7/16	14 5/8	8 3/4	63 3/8	111	5,210	469,500	709,500
2 1/2	15	9	65	107	5,528	492,000	744,000
2 9/16	15 3/8	9 1/4	66 5/8	105	5,810	516,000	778,500
2 5/8	15 3/4	9 7/16	68 1/4	103	6,105	540,000	813,000
2 11/16	16 1/8	9 11/16	69 7/8	99	6,410	565,000	849,000
2 3/4	16 1/2	9 7/8	71 1/2	97	6,725	590,000	885,000
2 13/16	16 7/8	10 1/8	73 1/8	95	7,040	615,000	925,000
2 7/8	17 1/4	10 3/8	74 3/4	93	7,365	640,000	965,000
2 15/16	17 5/8	10 5/8	76 3/8	91	7,696	666,500	1,005,000
3	18	10 13/16	78	89	8,035	693,000	1,045,000
3 1/16	18 3/8	11	79 5/8	87	8,379	720,500	1,086,500
3 1/8	18 3/4	11 1/4	81 1/4	85	8,736	748,000	1,128,000
3 3/16	19 1/8	11 1/2	82 7/8	85	9,093	776,050	1,169,000
3 1/4	19 1/2	11 11/16	84 1/2	83	9,460	804,100	1,210,000
3 5/16	19 7/8	11 15/16	86 1/8	81	9,828	833,150	1,253,000
3 3/8	20 1/4	12 1/8	87 3/4	79	10,210	862,200	1,296,000
3 7/16	20 5/8	12 3/8	89 3/8	77	10,599	892,100	1,339,550
3 1/2	21	12 5/8	91	77	10,998	922,000	1,383,100
3 5/8	21 3/4	12 15/16	94 1/4	73	11,607	1,021,000	1,566,000
3 3/4	22 1/2	13 3/8	97 1/2	71	12,626	1,120,000	1,750,000

All dimensions in inches. To convert to millimeters, multiply inches by 25.4.

Figure 29. Di-Lok Chain Dimensions and Test Requirements

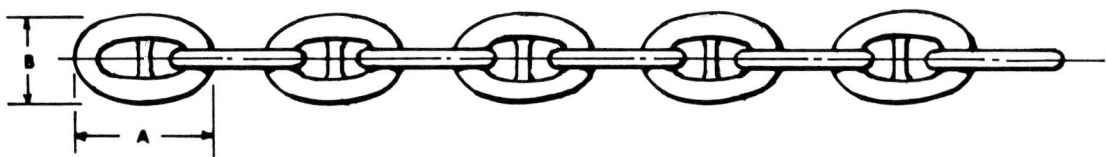

| CHAIN SIZE | DIMENSIONS |||| WEIGHT PER 15 FMSHOT (Approx) | TEST REQUIREMENTS |||||| NO. OF LINKS PER 15-FATHOM SHOT |
| | LINK LENGTH "A" | LINK WIDTH "B" | LENGTH OVER 5-LINKS "C" | GRIP RADIUS "D" || GRADE 1 || GRADE 2 || GRADE 3 |||
						PROOF LOAD	BREAK LOAD	PROOF LOAD	BREAK LOAD	PROOF LOAD	BREAK LOAD	
2	12	7-3/16	44	1-5/16	3360	159000	227000	227000	318000	318000	454000	133
2-1/16	12-3/8	7-7/16	45-3/8	1-3/8	3570	168500	241000	241000	337000	337000	482000	129
2-1/8	12-3/4	7-5/8	46-3/4	1-27/64	3790	178500	255000	255000	357000	357000	510000	125
2-3/16	13-1/8	7-7/8	48-1/8	1-15/32	4020	188500	269000	269000	377000	377000	538000	123
2-1/4	13-1/2	8-1/8	49-1/2	1-1/2	4250	198500	284000	284000	396000	396000	570000	119
2-5/16	13-7/8	8-5/16	50-7/8	1-17/32	4490	209000	299000	299000	418000	418000	598000	117
2-3/8	14-1/4	8-9/16	52-1/4	1-9/16	4730	212000	314000	314000	440000	440000	628000	113
2-7/16	14-5/8	8-3/4	53-5/8	1-5/8	4960	231000	330000	330000	462000	462000	660000	111
2-1/2	15	9	55	1-5/8	5270	242000	346000	346000	484000	484000	692000	107
2-9/16	15-3/8	9-1/4	56-3/8	1-11/16	5540	254000	363000	363000	507000	507000	726000	105
2-5/8	15-3/4	9-7/16	57-3/4	1-11/16	5820	265000	379000	379000	530000	530000	758000	103
2-11/16	16-1/8	9-11/16	59-1/8	1-3/4	6110	277000	396000	396000	554000	554000	792000	99
2-3/4	16-1/2	9-7/8	60-1/2	1-13/16	6410	289000	413000	413000	578000	578000	826000	97
2-13/16	16-7/8	10-1/8	61-7/8	1-27/32	6710	301000	431000	431000	603000	603000	861000	95
2-7/8	17-1/4	10-3/8	63-1/4	1-7/8	7020	314000	449000	449000	628000	628000	897000	93
2-15/16	17-5/8	10-9/16	64-5/8	1-7/8	7330	327000	467000	467000	654000	654000	934000	91
3	18	10-13/16	66	2	7650	340000	485000	485000	679000	679000	970000	89
3-1/16	18-3/8	11	67-3/8	2	7980	353000	504000	504000	705000	705000	1008000	87
3-1/8	18-3/4	11-1/4	68-3/4	2-1/16	8320	366000	523000	523000	732000	732000	1046000	85
3-3/16	19-1/8	11-1/2	70-1/8	2-1/16	8660	380000	542000	542000	759000	759000	1084000	85
3-1/4	19-1/2	11-11/16	71-1/2	2-1/8	9010	393000	562000	562000	787000	787000	1124000	83
3-5/16	19-7/8	11 15/16	72-7/8	2-1/8	9360	407000	582000	582000	814000	814000	1163000	81
3-3/8	20-1/4	12-1/8	74-1/4	2-3/16	9730	421000	602000	602000	843000	843000	1204000	79
3-7/16	20-5/8	12-3/8	75-5/8	2-3/16	10100	435000	622000	622000	871000	871000	1244000	77
3-1/2	21	12-5/8	77	2-5/16	10500	450000	643000	643000	900000	900000	1285000	77
3-5/8	21-3/4	12-15/16	79-3/4	2-5/16	11300	479000	685000	685000	958000	958000	1369000	73
3-3/4	22-1/2	13-3/8	82-1/2	2-15/32	12000	509000	728000	728000	1019000	1019000	1455000	71
3-7/8	23-1/4	14	85-1/4	2-15/32	12900	540000	772000	772000	1080000	1080000	1543000	69

Figure 30. Flash-butt Welded Chain Dimensions and Test Requirements

WIRE ROPE

The advantage of using wire rope in mooring lines is its high strength-to-weight ratio relative to chain. However, several disadvantages exist. Careful handling is necessary to avoid kinks and wear by abrasion. Kinks, abrasions, and corrosion can significantly reduce the strength of a wire rope. Corrosion can be minimized by galvanizing a wire rope; however, this increases cost about 20% and decreases strength about 10%. If a wire rope mooring line parts near the middle, it is normally removed from service, whereas a chain can be salvaged and repaired. Also, wire rope is usually stored on winch drums on the deck of a vessel, which increases deck load and decreases stability of the vessel when on a rig move. By comparison, anchor chain is stored in chain lockers below deck, thereby enhancing stability. Finally, the pulling power of a winch decreases with the amount of rope on the drum because successive rope layers are higher from the axis of the drum.

Wire rope is manufactured, as shown in figure 31, by first winding individual wires into strands and then winding the strands together around a core. Figure 32 shows cross

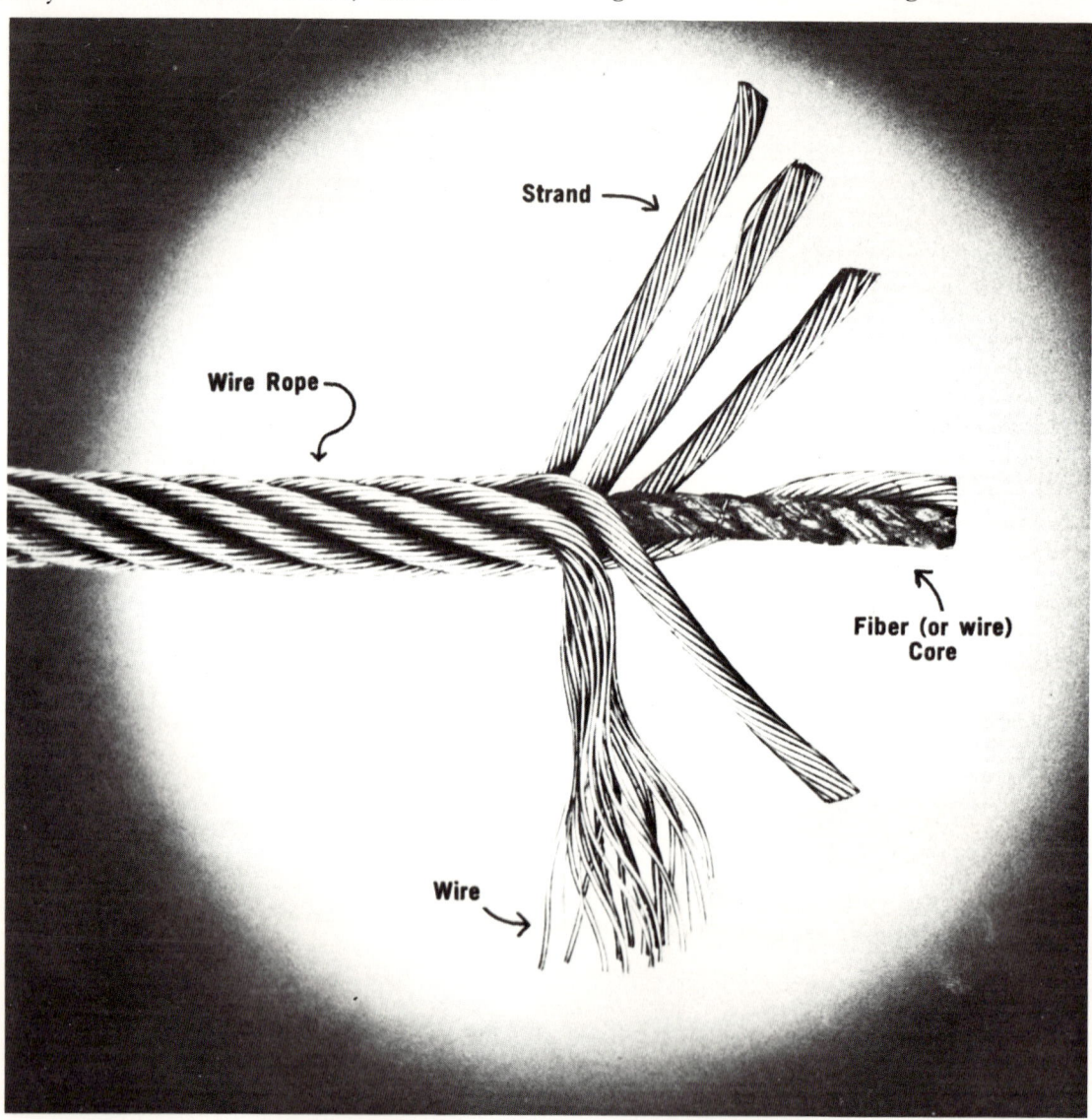

Figure 31. Wire Rope Fabrication Nomenclature

sections of some typical wire ropes. Almost all wire rope used in mooring systems is of either the 6 x 19 or the 6 x 37 class. The number 6 refers to the number of strands per rope and the number 19 or 37 refers to the number of individual wires per strand. The number of wires per strand can vary within each class in the illustration. The 6x19 class has relatively large individual wires; therefore, it has high resistance to corrosion and wear by abrasion,

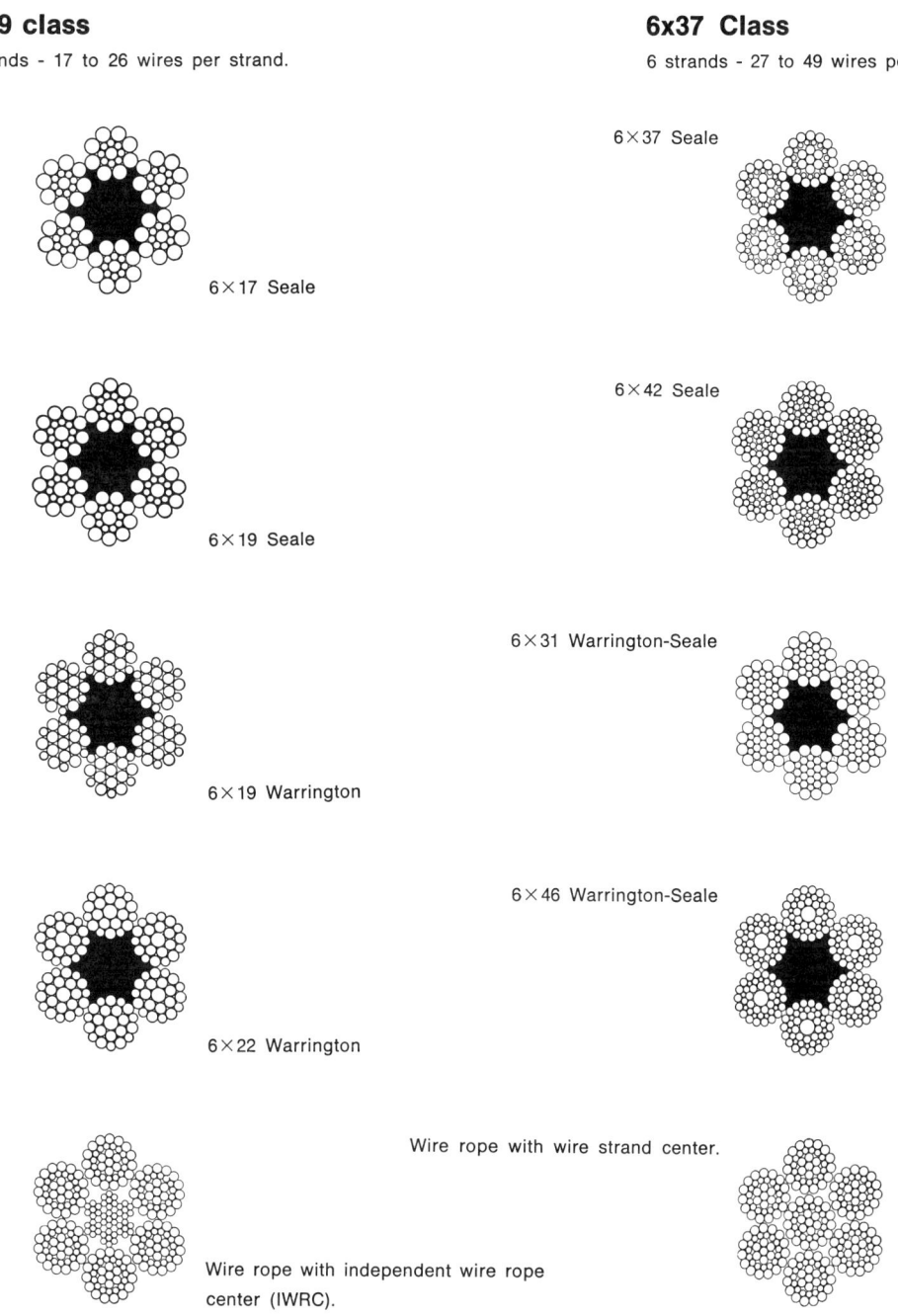

Figure 32. Classification of Wire Rope

along with good flexibility and fatigue life. Smaller individual wires in the 6 x 37 class result in high flexibility and excellent fatigue life; however, corrosion and abrasion resistance is poorer. With these considerations, choice of a wire rope size depends on the particular application. Mooring lines, which may be two and one-half to three and one-half inches in diameter, are usually 6 x 37 class rope for flexibility and are galvanized for corrosion protection. Pendant lines are subjected to severe abuse when anchors are run and pulled. Therefore, 6x19 wire rope is usually used for this service to withstand abrasion and mechanical damage. Figure 33 lists strengths for various sizes and types of galvanized wire rope.

The two most common types of wire rope construction are Warrington and Seale. The Warrington construction has both large and small wires in the outer layer of each strand, offering good flexibility and fatigue life, but maintaining poor resistance to abrasion and corrosion. Abrasion and corrosion resistance is improved in the Seale construction where an outer layer of large wires protects an inner layer of small wires. Usually, the Seale construction is specified for mooring applications.

Most of the wire rope cross-sections shown previously are constructed with fiber cores, which make them unacceptable for moorings where wire ropes must tolerate considerable abuse by bending over small radii, crushing on drums, and rubbing on sheaves. Only ropes with metallic cores should be used. Two types are available, the independent wire rope core (IWRC) and the wire strand core (WSC). Generally, IWRC is preferred, inasmuch as WSC causes the rope to be much stiffer. IWRC increases rope weight about 10% and breaking strength about 7%, and increases resistance to crushing and abrasion.

The way in which the wires and strands are wound together is called the "lay" of a rope. Right lay and left lay refer to the appearance of a rope to an observer as he looks along the rope axis from above. If the strands twist to the right, the rope is referred to as right lay, or standard lay. If they turn to the left, the rope is left lay.

Regular lay refers to ropes where the twist of the wires in the strands is opposite to the twist of the strands in the rope. Therefore, right regular lay, which is standard, indicates that the strands twist to the right and the individual wires in the strands twist to the left. In lang lay ropes, the wires and strands twist in the same direction. The outer wires of regular lay ropes run nearly parallel to the axis of the rope, whereas in lang lay ropes they run diagonally across the axis. Because the wires and strands of regular lay ropes are twisted in opposite directions, the tendency to kink and untwist is much less than for lang lay ropes. Also, lang lay ropes are more susceptible to crushing abuse.

Wire rope should be lubricated frequently to prolong service life. S. M. Acaster (1972) notes that variations in tension will cause a wire rope to "work" and pump water into the core. Therefore, wire rope lubricants should be of a penetrating type which will protect the smaller inside wires. Heavy, bitumastic-type lubricants may tend to harden and seal water inside, resulting in severe internal corrosion. Practice has shown this type of wire rope lubrication to be generally unfavorable.

Two grades of carbon steel are commonly used in manufacturing wire rope, Improved Plow Steel (IPS) and Extra-Improved Plow Steel (XIPS). XIPS has superior breaking strength and abrasion resistance; however, its fatigue resistance is normally poorer than IPS.

As stated earlier, wire rope must be handled carefully to avoid damage by abuse. Physical damage often occurs when wire rope is run over sheaves which are too small or in poor condition. Within the limits of practicality, sheave diameters should be as large as possible. Ideally, sheave diameters in the range of 30 to 50 times the wire rope diameter are suggested; however, this may be too large for many applications. Wire rope should be coiled on winch drums neatly, with the rope resting in the valleys of the coil below. If the rope cuts across an adjacent coil, severe crushing can occur when the line is subjected to high tension.

REVIEW QUESTIONS

LESSONS IN ROTARY DRILLING

Unit V, Lesson 2: Spread Mooring Systems

1. Ancient mooring systems used stones attached to _____ for the anchoring systems.

    a. chain
    b. wire rope
    c. fiber rope

2. Submersible barges were the first mobile drilling units but were restricted to about _____ feet of depth.

    a. 90
    b. 175
    c. 325

3. The maximum ball-joint angle allowable for drilling operations is _____ degrees, with maximum vessel offset being 6 percent.

    a. 10
    b. 4
    c. 25

4. An anchor does not set but is pulled up with a large ball of mud covering it. This condition is known as _____ _____.

    a. trip failure
    b. balling up
    c. anchor turn-over

5. For soft bottoms, anchor fluke angle should be 50 degrees. For hard bottoms, fluke angle should be _____ degrees.

6. Under the force of its own weight, the suspended length of a mooring line will fall into a shape known as a _____ _____.

7. When more than one mooring line is used, the environmental force is _____.

    a. neglected
    b. shared between them
    c. doubled

8. Because chain has a low strength-to-weight ratio, many deepwater mooring systems employ lighter, stronger _____ _____ as the mooring line.

9. Composite lines use both _____ and _____.

10. Name three different types of anchors used in spread mooring systems.

    (1) _____

    (2) _____

    (3) _____

11. All chain used in mooring patterns is of the _____ type, which prevents kinking and stops distortion.

12. In a 6 × 19 class wire rope, the *6* represents the _____ _____ and the *19* refers to the number of individual wires per strand.

13. Two types of wire rope are the IWRC and WSC. What do the letters represent, and which type is preferred?

    _____

    _____

    _____

14. The way in which the wires of a rope are wound together is known as the _____ of a rope.

15. A pelican hook chain stopper is commonly used on anchor handling boats to cut pendant lines and anchor chains. The advantage of this type stopper is that it can be _____ when under tension.

16. Wire rope is stored on _____.

17. Windlasses are used to pull in and run out _____.

18. Pendant lines are used to pull and lower the anchors, and their ends are held at the water surface by _____ _____.

19. If an anchor will not hold when preloaded, installation of a second anchor on one mooring line, or what is known as "_____," may be necessary.

20. Mooring can be complicated because as many as _____ different companies or organizations may be involved.

TABLE 3.8
6 x 19 CLASSIFICATION WIRE ROPE, GALVANIZED,
INDEPENDENT WIRE ROPE CORE

1	2	3	4	5	6	7	8	9	10
\multicolumn{2}{c	}{Nominal Diameter}	\multicolumn{2}{c	}{Approx. Mass}	\multicolumn{6}{c}{Nominal Strength}
				\multicolumn{3}{c	}{Improved Plow Steel}	\multicolumn{3}{c}{Extra Improved Plow Steel}
in.	mm	lb/ft	kg/m	lb	kN	Metric Tonnes	lb	kN	Metric Tonnes
½	13	0.46	0.68	20,600	91.6	9.34	23,940	106	10.9
9/16	14.5	0.59	0.88	26,000	116	11.8	30,240	135	13.7
5/8	16	0.72	1.07	32,200	143	14.6	37,080	165	16.8
¾	19	1.04	1.55	46,000	205	20.9	52,920	235	24.0
7/8	22	1.42	2.11	62,200	277	28.2	71,640	319	32.5
1	26	1.85	2.75	80,800	359	36.7	93,060	414	42.2
1⅛	29	2.34	3.48	101,600	452	46.1	117,000	520	53.1
1¼	32	2.89	4.30	125,000	556	56.7	143,800	640	65.2
1⅜	35	3.50	5.21	150,200	668	68.1	172,800	769	78.4
1½	38	4.16	6.19	178,000	792	80.7	205,200	913	93.1
1⅝	42	4.88	7.26	207,000	921	93.9	237,600	1060	108
1¾	45	5.67	8.44	239,400	1060	109	275,400	1230	125
1⅞	48	6.50	9.67	273,600	1220	124	313,200	1390	142
2	51	7.39	11.0	309,600	1380	140	356,400	1590	162

TABLE 3.9
6 x 37 CLASSIFICATION WIRE ROPE, BRIGHT (UNCOATED) OR
DRAWN-GALVANIZED WIRE, INDEPENDENT
WIRE-ROPE CORE

1	2	3	4	5	6	7	8	9	10
\multicolumn{2}{c	}{Nominal Diameter}	\multicolumn{2}{c	}{Approx. Mass}	\multicolumn{6}{c}{Nominal Strength}
				\multicolumn{3}{c	}{Improved Plow Steel}	\multicolumn{3}{c}{Extra Improved Plow Steel}
in.	mm	lb/ft	kg/m	lb	kN	Metric Tonnes	lb	kN	Metric Tonnes
½	13	0.46	0.68	23,000	102	10.4	26,600	118	12.1
9/16	14.5	0.59	0.88	29,000	129	13.2	33,600	149	15.2
5/8	16	0.72	1.07	35,800	159	16.2	41,200	183	18.7
¾	19	1.04	1.55	51,200	228	23.2	58,800	262	26.7
7/8	22	1.42	2.11	69,200	308	31.4	79,600	354	36.1
1	26	1.85	2.75	89,800	399	40.7	103,400	460	46.9
1⅛	29	2.34	3.48	113,000	503	51.3	130,000	578	59.0
1¼	32	2.89	4.30	138,800	617	63.0	159,800	711	72.5
1⅜	35	3.50	5.21	167,000	743	75.7	192,000	854	87.1
1½	38	4.16	6.19	197,800	880	89.7	228,000	1010	103
1⅝	42	4.88	7.26	230,000	1020	104	264,000	1170	120
1¾	45	5.67	8.44	266,000	1180	121	306,000	1360	139
1⅞	48	6.50	9.67	304,000	1350	138	348,000	1550	158
2	51	7.39	11.0	344,000	1530	156	396,000	1760	180
2⅛	54	8.35	12.4	384,000	1710	174	442,000	1970	200
2¼	57	9.36	13.9	430,000	1910	195	494,000	2200	224
2⅜	61	10.4	15.5	478,000	2130	217	548,000	2440	249
2½	64	11.6	17.3	524,000	2330	238	604,000	2690	274
2⅝	67	12.8	19.0	576,000	2560	261	662,000	2940	300
2¾	70	14.0	20.8	628,000	2790	285	722,000	3210	327
2⅞	74	15.3	22.8	682,000	3030	309	784,000	3490	356
3	77	16.6	24.7	740,000	3290	336	850,000	3780	386
3⅛	80	18.0	26.8	798,000	3550	362	916,000	4070	415
3¼	83	19.5	29.0	858,000	3820	389	984,000	4380	446
3⅜	86	21.0	31.3	918,000	4080	416	1,058,000	4710	480
3½	90	22.7	33.8	982,000	4370	445	1,128,000	5020	512
3¾	96	26.0	38.7	1,114,000	4960	505	1,282,000	5700	582
4	103	29.6	44.0	1,254,000	5580	569	1,440,000	6410	653

Figure 33. Strengths of Galvanized Wire Ropes (API Spec. 9A)

CONNECTING ELEMENTS

Generally, connecting elements represent weak points in a mooring system. In a chain system, failures occur in shackles, swivels, or detachable links. This is especially true with fatigue failures, where life expectancies of connecting elements may be only 30% to 50% of the chain itself. Connecting elements should be fabricated from forged steel, not cast steel, and should be inspected and tested frequently.

Wire ropes tend to kink and fatigue near the end fittings. To prevent premature failures, a good policy is to "cut back" wire rope by approximately 15 feet at end fittings as often as every rig move.

Chain Fittings

Various types of chain fittings are shown in figures 34 through 39. The Kenter connecting link is often preferred to the Baldt detachable link primarily because of its longer fatigue life. The Kenter is somewhat larger, however, and may not fit certain types of chain handling equipment. Some operators prefer not to use

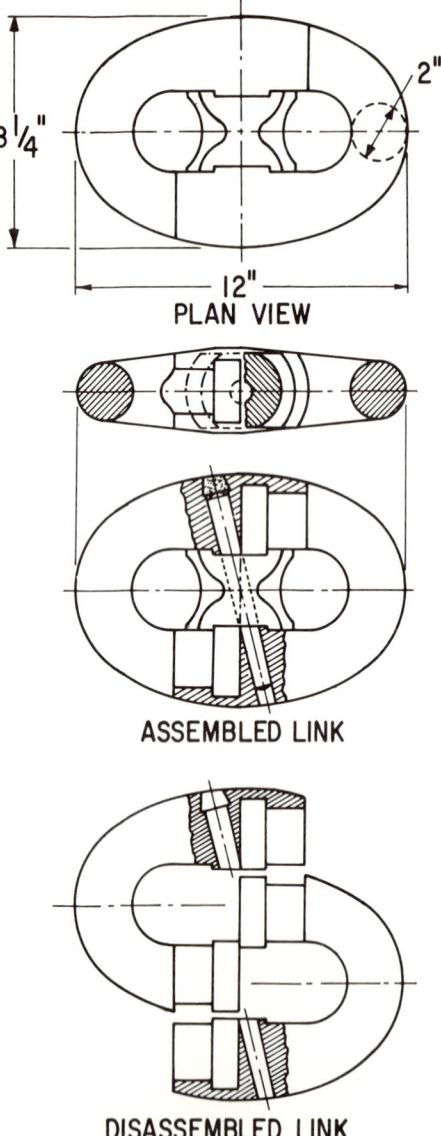

Figure 34. Kenter Connecting Link

swivels in a mooring line. Experience indicates that swivels have a short fatigue life and the last threads tend to be troublesome weak points. A pelican hook chain stopper (figure 38) is commonly used on anchor handling boats to snub off pendant lines and anchor chains. The advantage of the pelican hook is that it can be released when under tension. An Ulster pawl-type stopper is shown in figure 39.

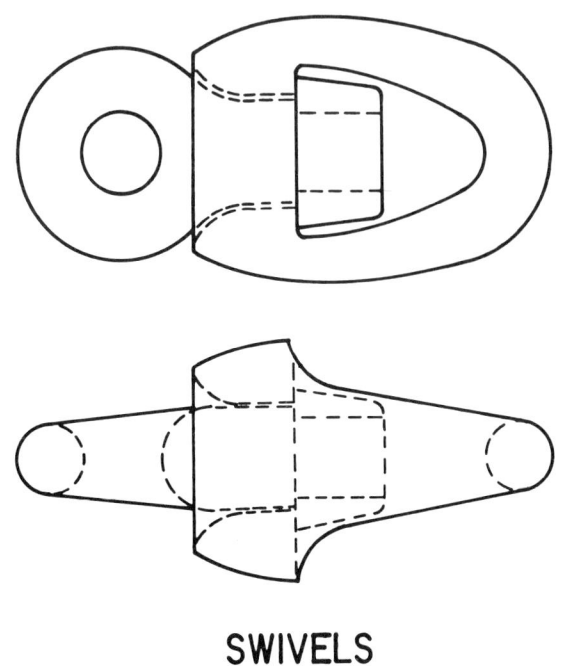

SWIVELS

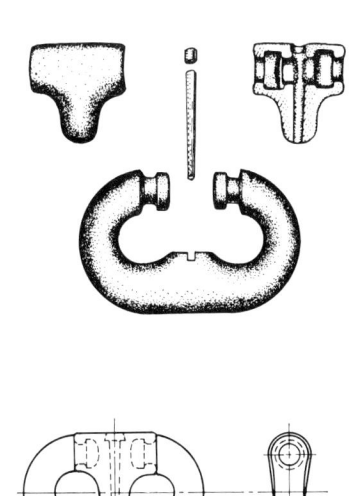

Figure 35. Baldt Detachable Chain Connecting Link

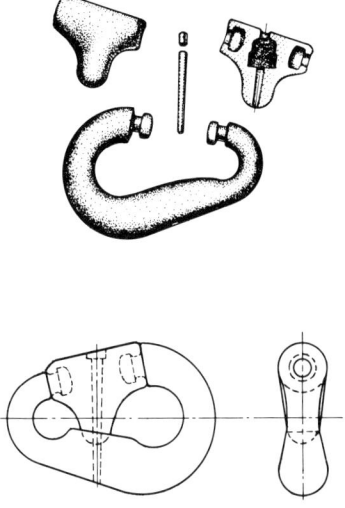

Figure 36. Baldt Detachable Anchor Connecting Link

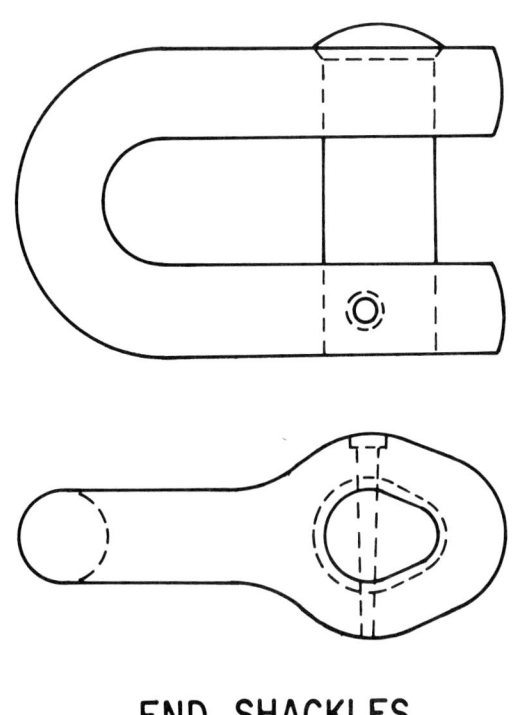

END SHACKLES

Figure 37. Baldt Swivels and End Shackles

25

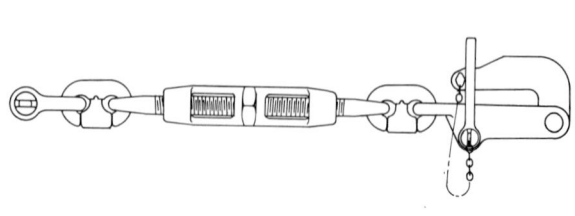

Figure 38. Pelican Hook Type Chain Stopper

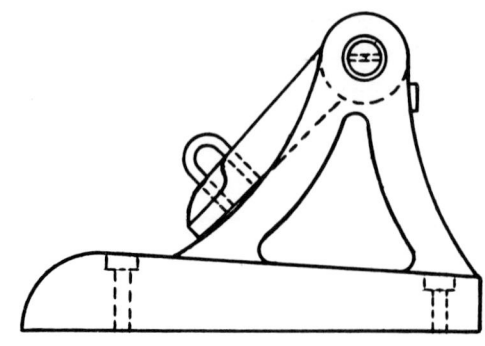

Figure 39. Ulster (Pawl) Type Chain Stopper

Wire Rope Fittings

The end fittings of wire rope must be manufactured properly in order to develop the full strength of a mooring system. Two common types of socket fittings are the swaged socket and the zinc-poured socket (see figure 40). Swaged sockets are attached to the rope by inserting the rope into the shank hole and pressing or swaging the shank onto the end of the rope with a special tool. To attach zinc-poured sockets, the end of the rope is frayed, or "broomed out," to prevent the strands from untwisting. The broomed-out end is then inserted into the socket and the socket bowl is filled with molten zinc.

Full holding power is sensitive to extreme care in the installation procedure. As a result, some contractors prefer swaged sockets in view of their greater simplicity. Others prefer zinc-poured sockets because they lend themselves more readily to field repairs. Regardless, all sockets should be made from forged steel rather than cast steel in order to provide extra strength in heavy-duty service.

Wire rope mooring lines are sometimes fitted with eyes and thimbles (figure 41) rather than sockets. The best type of thimble has a plate insert to protect it from crushing when under a heavy load.

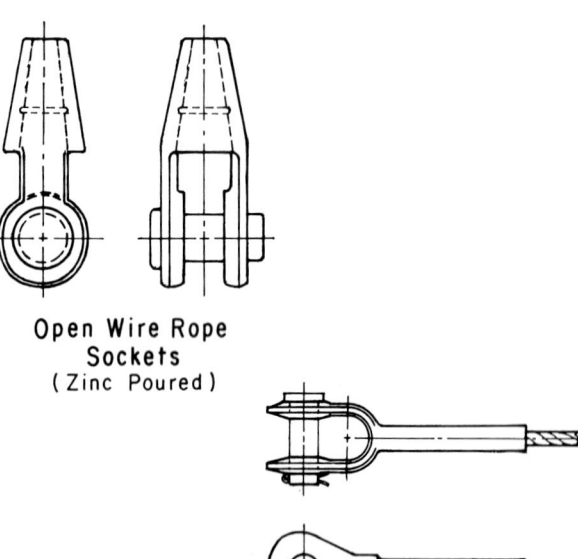

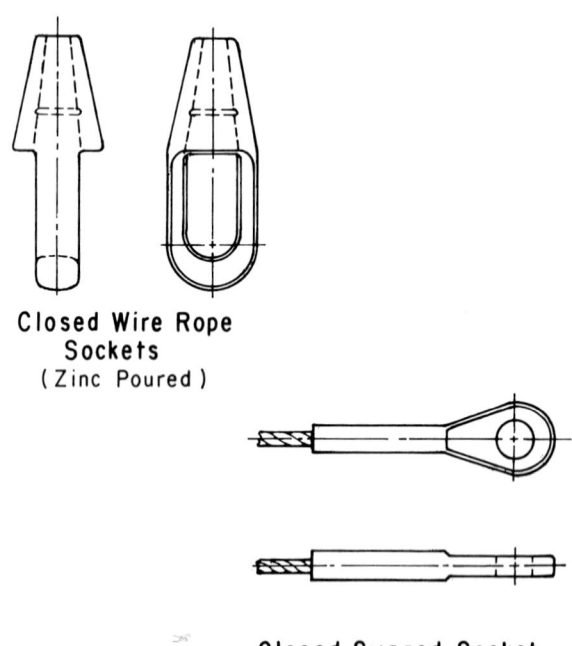

Figure 40. Wire Rope Fittings

26

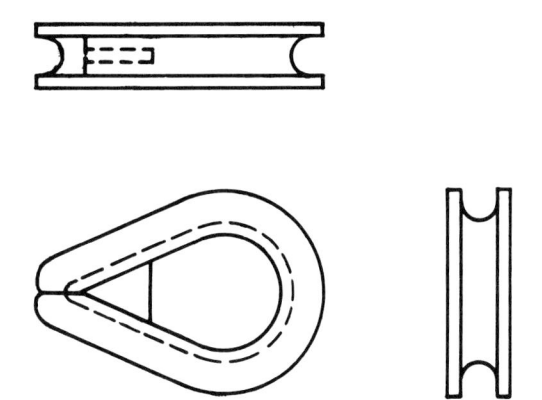

Figure 41. Wire Rope Thimble

WINCHES

Winches are used to handle and store wire rope. Dual drum winches (figure 42) are most common; however, single drum and quadruple drum units have been used. The size of a winch depends on the amount of wire rope to be stored on the drum and the maximum line pull to be exerted. A winch should be able to pull half the breaking strength of a mooring line, and should be equipped with mechanical brakes which can hold full breaking strength. Anchor winches on a drilling rig are usually driven by direct current (DC) electrical motors which are powered by the rig engines. Individual engines for the winches are too expensive, and under blowout conditions, a minimum number of engines should be running in order to reduce spark and fire hazards.

In case a blowout occurs, emergency release systems should be built into the mooring system so that the vessel can move off location quickly. Even if the blowout does not ignite, gas content in the water can reduce buoyancy to the point where a vessel may capsize and be lost. Usually, the last layer of wire rope on a winch drum is not paid out; therefore, the rope end can be loosely clamped in place. Should an emergency occur, the clamp can be knocked away quickly. An alternate method of freeing a mooring line is to position a hydraulic cable cutter below the water surface where sparks will not occur, lessening the chances for an explosion.

It is important to wind rope onto the winch drum properly. Correct winding depends on

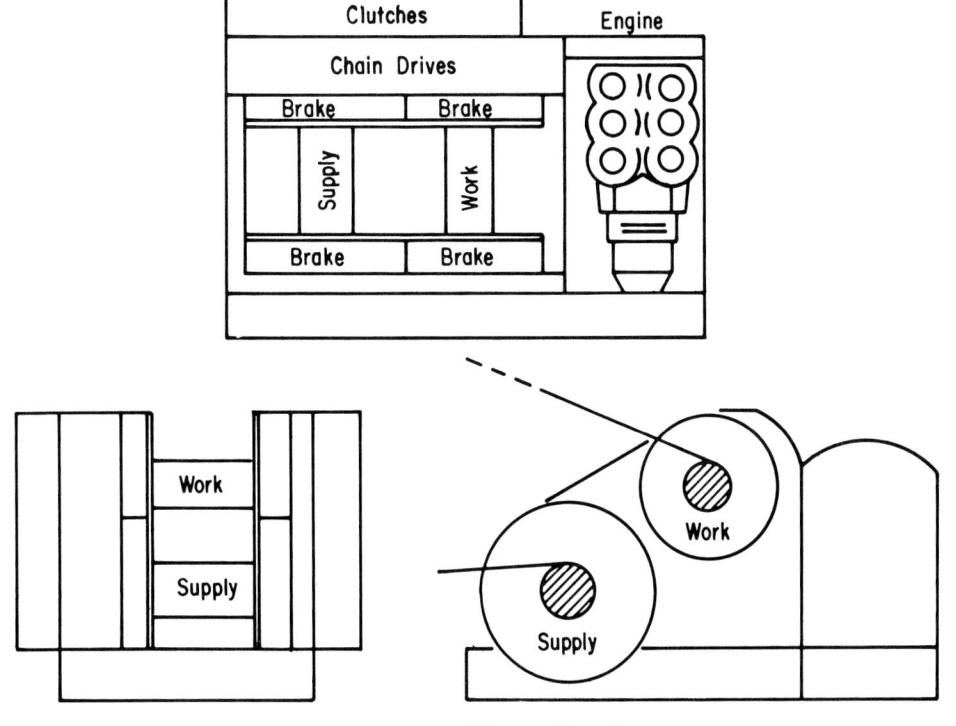

Figure 42. Dual Drum Winch

27

the direction of winding and the lay of the rope, as outlined in figure 43. Ideally, the winch drum should be grooved and the rope wound so that it does not cut across adjacent layers. This will minimize crushing and wear, and will extend the service life of the rope.

WINDLASSES

Windlasses are used to pull in and run out chain. The chain is not stored on the windlass itself, but in chain lockers below. To propel the chain, windlasses are fitted with special gear-like sheaves, known as "wildcats." The wildcats are built with sprocket teeth, called "whelps," and pockets, which engage individual chain links and hold them securely. Figure 44 shows a windlass wildcat and part of the deck equipment used for handling chain. Chain may also be passed over a grooved sheave, also shown in figure 44. Tension forces on the chain may bend the links lying flat against the sheave. The level of the bending stresses created will depend on the magnitude and direction of the tension forces in the chain.

Ahead of the windlass is a chain stopper which holds the entire tension of the mooring line when the windlass is not in use. To handle emergency situations such as blowouts, the design of the chain stopper should allow for quick release without supplying power to the windlass. Also, the end of the chain should be left free in the chain locker, or an extra length of chain should be supplied to allow the rig to move off location and away from danger should it occur.

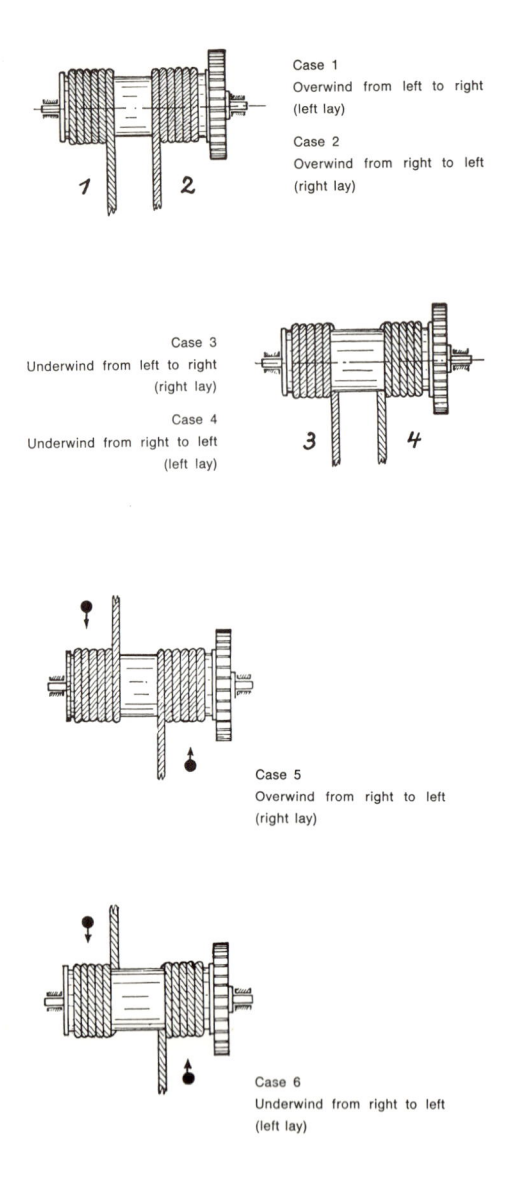

Case 1
Overwind from left to right
(left lay)

Case 2
Overwind from right to left
(right lay)

Case 3
Underwind from left to right
(right lay)

Case 4
Underwind from right to left
(left lay)

Case 5
Overwind from right to left
(right lay)

Case 6
Underwind from right to left
(left lay)

Figure 43. Wire Rope Winding Procedures

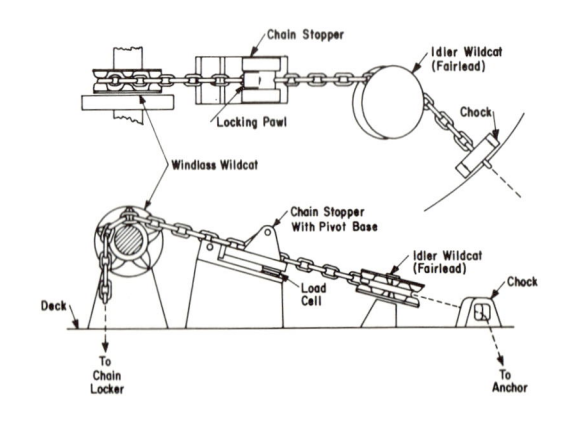

Figure 44. Windlass Wildcat and Chain Handling Equipment

A good fit between the chain and the whelps of the wildcat is imporant to avoid damaging the windlass. Figure 45 illustrates one techinque for checking the fit and determining any needed modifications.

Size requirements for windlasses are similar to those for winches. A windlass should be able to pull approximately one-half the rated breaking strength of the chain passing over it. Typically, windlasses are usually powered by DC electric motors driven by the rig engines, although some vessels use independent engine drives.

MOORING LINE TENSION MEASURING DEVICES

In rough weather particularly, mooring line tensions must be monitored carefully. If tensions greater than one-third breaking strength occur in the windward lines, it may be necessary to slack the leeward lines. Proper decisions depend on accurate measurement of the mooring line tensions. Measurement procedures and equipment differ for mooring lines made of wire rope as opposed to chain.

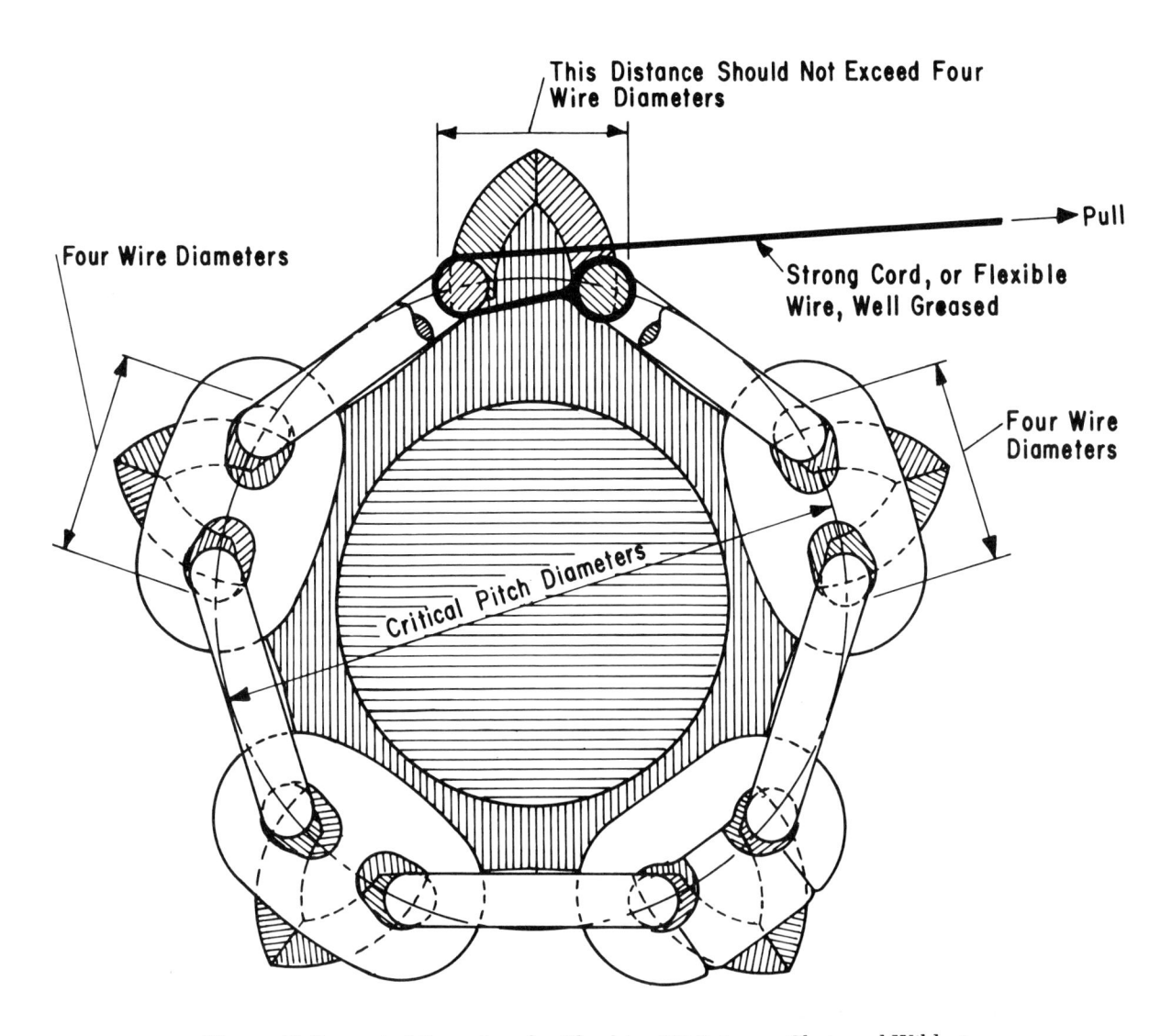

Figure 45. Suggested Procedure for Checking Fit Between Chain and Wildcat

Chain Tension Measurements

Most techniques for measuring chain tensions are based on the physical principle of balancing forces and lever arms. An example is shown in figure 46, where the chain stopper is mounted on a pivot. The chain tension is eccentric to the pivot, acting through a lever arm "A." An overturning moment on the chain stopper results, which is counterbalanced by the force in the load cell acting through the lever arm "B." The load cell force can be measured and multiplied by the ratio B/A to give the chain tension. This tension measuring system is inferior, however, because considerable tension may be absorbed by the fairlead and the chock, especially the chock. As a result, mooring line tension beyond the chock may be greater than that indicated by the load cell.

One of several other approaches to measuring chain tension is to mount the entire windlass and chain stopper assembly on a pivot and load cell, as shown in figure 47. The chain leads downward to a swivel fairlead. If the fairlead has minimum friction, chain tension indicated by the load cell should be fairly accurate. A swivel fairlead is shown in figure 48.

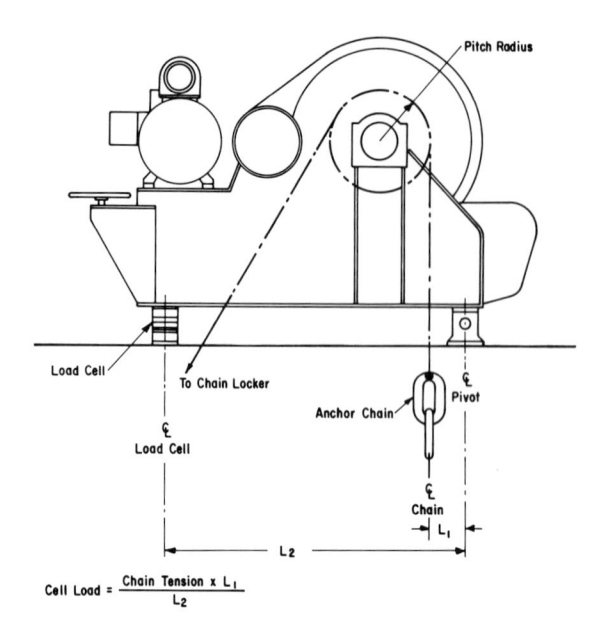

$$\text{Cell Load} = \frac{\text{Chain Tension} \times L_1}{L_2}$$

Figure 47. Another Arrangement to Measure Chain Line Tension

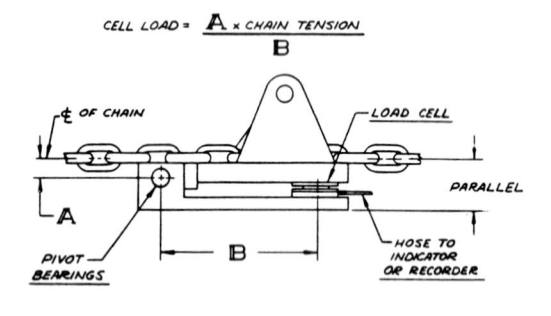

$$\text{CELL LOAD} = \frac{A \times \text{CHAIN TENSION}}{B}$$

Figure 46. One Arrangement for Measuring Chain Line Tension

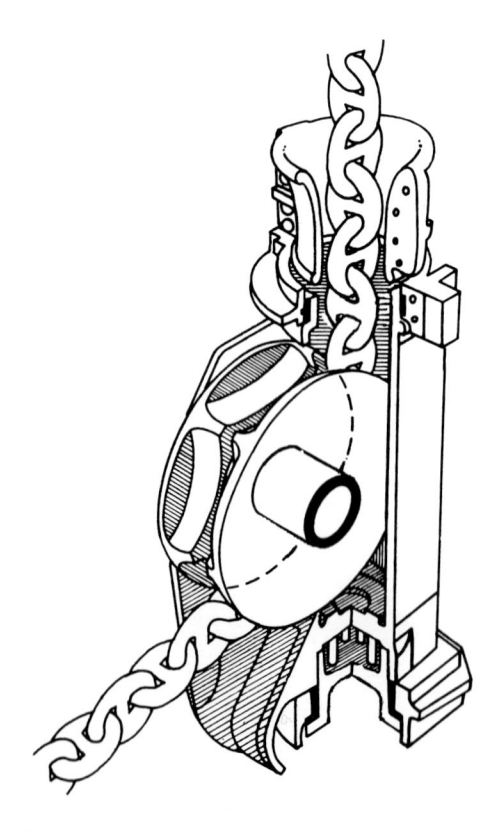

Figure 48. Swivel Fairlead for Anchor Chain

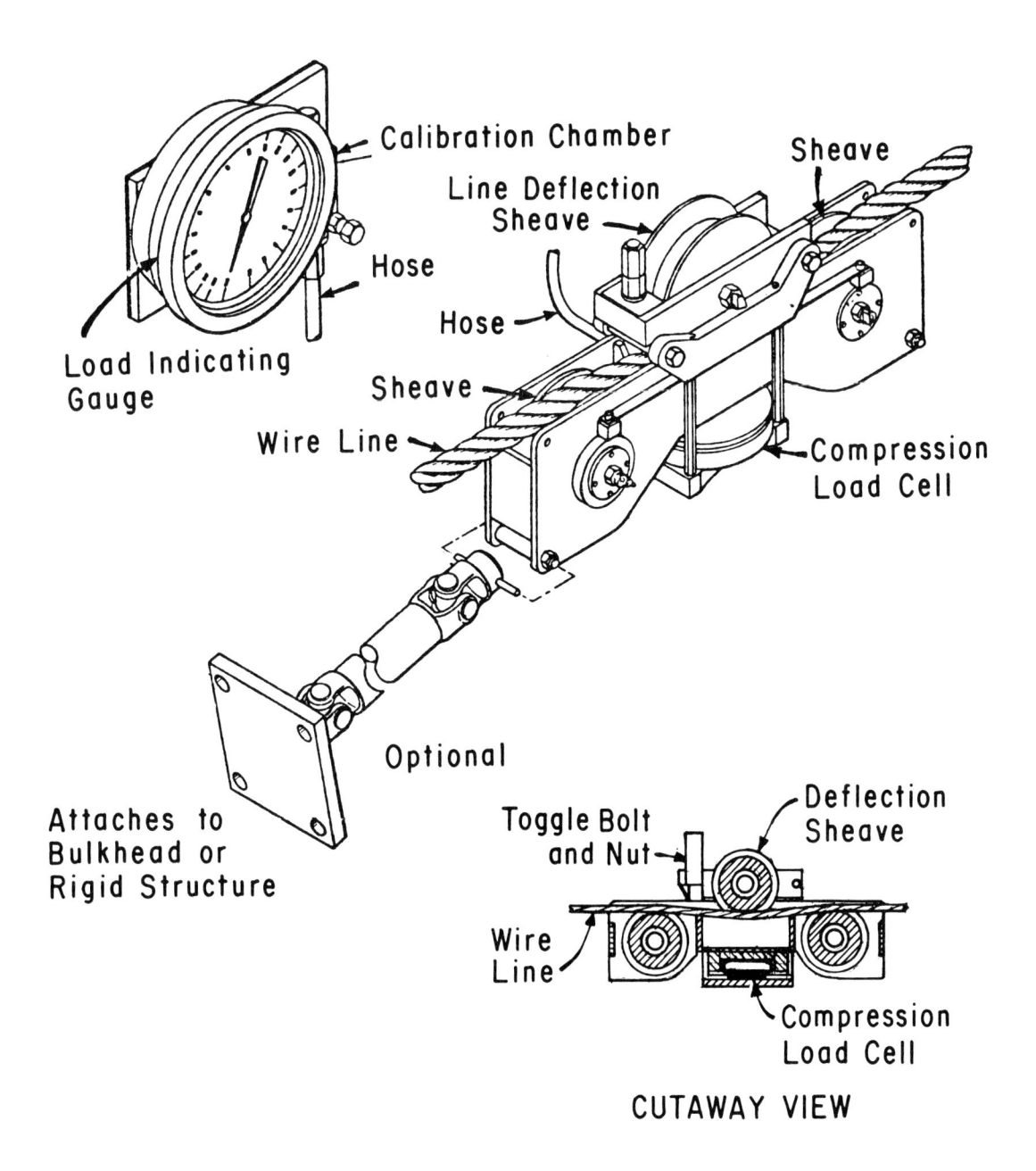

Figure 49. Three-Sheave Wire Line Tension Measuring Device

Wire Rope Tension Measurements

Wire rope tensions are commonly determined by measuring the force required to deflect the wire rope from a straight line. A three-sheave tension measuring device which incorporates this basic principle is shown in figure 49. The deflection angle of the line and the force on the deflection sheave are uniquely related. The two outboard sheaves and the deflection sheave are designed to maintain a specified included angle, thereby fixing the geometry of the system. With the geometry fixed, the load cell reading can be calibrated to yield the wire line tension directly.

PENDANT LINES AND MOORING BUOYS

Pendant lines are used to pull and lower the anchors, and their ends are held at the water surface by mooring buoys. Generally, the diameter of pendant lines varies between one and three-quarter and two and one-quarter inches. It is important that the pendant lines have the proper length. If a pendant line is too short, the mooring buoy may be pulled under and damaged in heavy seas. If too long, the excess line may pile up on bottom and become tangled in the anchor, particularly in hard bottom. Common practice is to make the length of pendant lines 50 to 100 feet greater than the water depth. In shallow water, the extra length may be taken as a percentage of water depth, up to 25% in some cases.

The tendency of the pendant line to tangle with the anchor is countered in various ways. One technique is to place a length of chain between the crown of the anchor and the lower end of the pendant line. This is called a crown chain. Another technique is to hold the lower end of the pendant line taut by means of a "spring buoy," similar to the arrangement shown in figure 50.

Considerable attention must be given to the design details of mooring buoys. They should be floated broadside down to minimize "bobbing" behavior, and should be designed to reduce movement in the propellor wash of the anchor handling boat. To haul on board the anchor handling boat, buoys should be equipped with a cleat or cross on top by which they can be "lassoed" with a loop of rope. The cross and the connection to the pendant line at the bottom should be joined by a strength member for lifting. The shell of a mooring buoy is usually steel, and the interior is normally filled with polyurethane foam to prevent sinking in the event of puncture. Mooring buoys manufactured from synthetic materials such as rubber and plastics have been marketed; however, they are expensive and may not be sufficiently durable for extended service. In some areas, mooring buoys can be hazards to navigation, and legal regulations may require that they be equipped with lights, radar reflectors, and other aids.

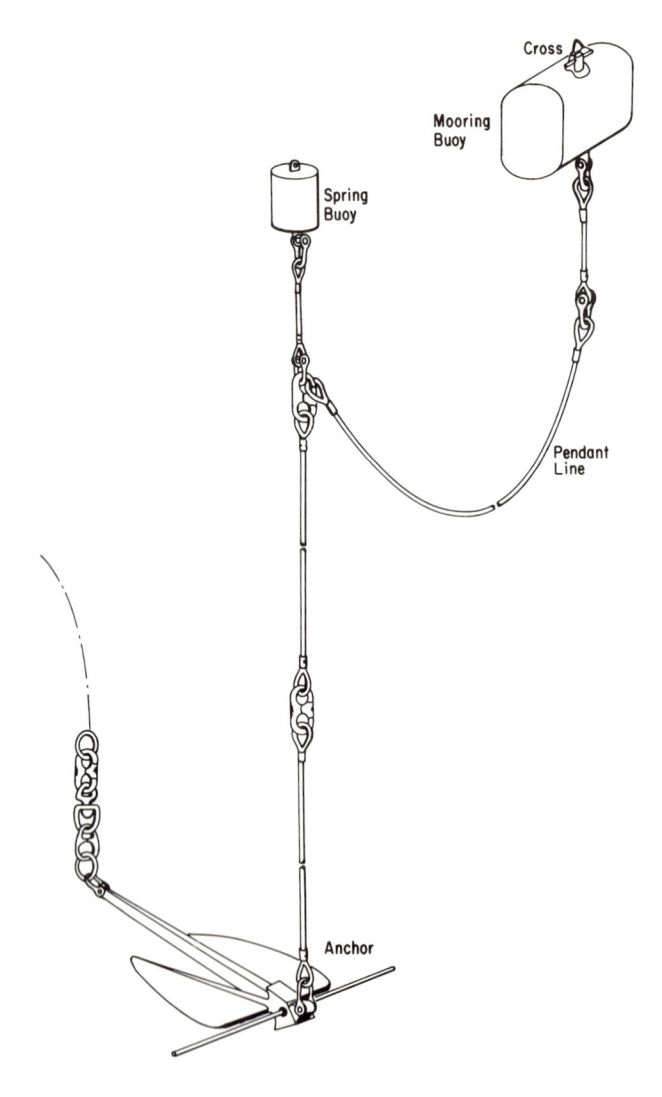

Figure 50. Spring Buoy to Hold Lower End of Pendant Line Taut

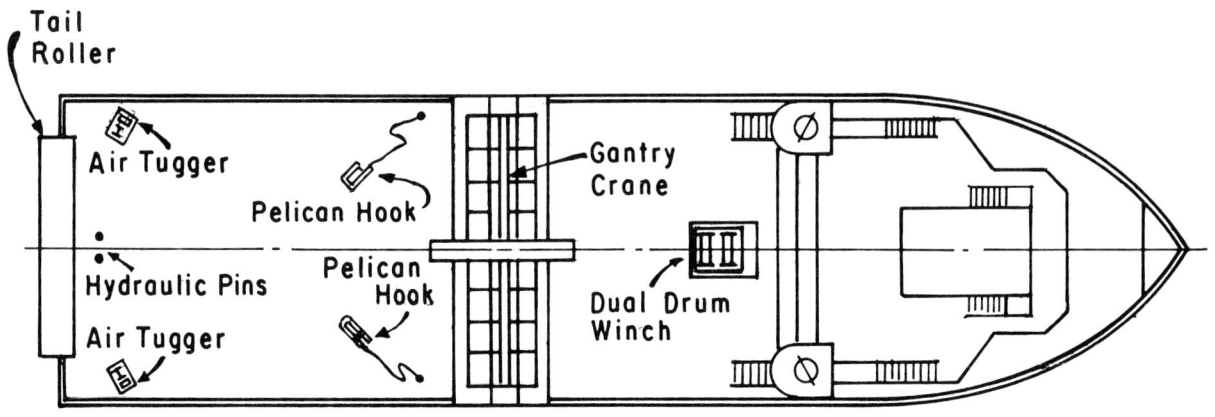

Figure 51. Deck Equipment Arrangement for Anchor Handling

ANCHOR HANDLING BOATS

Although anchor handling boats are not an integral part of the mooring system itself, they are essential for installation and recovery operations. Primarily, they must be able to exert sufficient pull to run out the mooring lines to design lengths. This is a particular problem with chain because of its weight and a tendency for it to drag on bottom. Bottom drag friction creates considerable resistance which the anchor handling boat must overcome. Pull of an anchor handling boat is primarily a function of the continuous shaft horsepower (SHP) available to the propellers. This is distinct from installed horsepower and input horsepower, which have different meanings. The amount of pull per SHP can be improved by the use of controllable pitch propellers, "Kort" nozzles (airfoil-shaped rings surrounding the propellers), and efficient hull design.

Anchor handling boats should have watertight compartments. Anchors have been pulled into the bottom of the boat causing the boat to be holed, flooded, and sunk. Also, anchor handling boats are sometimes subjected to severe abuse, and hull plates may crack and leak under prolonged service.

The deck area should be outfitted and arranged somewhat as shown in figure 51. When an anchor is being pulled or set, the weight of the anchor is held on the pendant line which is passed over the tail roller, between the hydraulic pins, and onto the winch drum. Avoiding small-radius bending damage to the pendant line requires the diameter of the tail roller to be large. A four-foot diameter is minimum, and a six-foot diameter is preferable. Hydraulic pins are provided to keep the pendant line centered on the tail roller and are retractable when not needed. Additional techniques for centering the pendant line are to make the tail roller into the shape of an hourglass and/or to form a groove into the center.

The winch should have 200,000 lbs. to 300,000 lbs. pulling capacity, and should have mechanical brakes on the drum. To release the pendant line from the winch and attach a buoy, for example, the pendant line must be snubbed off with the pelican hooks. Air tuggers are positioned at convenient locations to handle heavy objects such as buoys when they are on deck.

The gantry crane is used to handle heavy objects over the tail roller, and should be moved forward when not in use. Instead of a travelling gantry crane, many anchor handling vessels are equipped with fixed A-frames, or post hoists, at the stern. These can be troublesome when mooring floating drilling vessels in rough weather. For example, when the anchor handling boat is in close proximity to the rig stern-first, the A-frame may catch under structural members, resulting in damage to both boat and rig. A vessel equipped with a gantry crane has the crane moved forward and out of the way to avoid such hazards.

CHAPTER III
PLACING AND RECOVERING MOORINGS

PLANNING AND ORGANIZATION

A successful mooring operation requires considerable advance planning and organization. The well site should be marked and surveyed, and the bottom conditions established before the rig moves onto location. With water depth known, pendant lines can be checked for correct lengths. All mooring components and handling equipment should be inventoried and inspected to assure that they are present and in good condition. Missing or damaged items can be replaced or repaired as needed. Equipment and procedures to be used should be reviewed with all concerned personnel to assure an efficient, safe operation.

Constant and close coordination between the drilling rig and the anchor handling boat is essential during the mooring operation. Clear lines of communication should be established beforehand to avoid any confusion while the job is in progress. This task is complicated by the fact that as many as four different companies or organizations may be involved. First is the operator, or oil company, for whom the well is drilled. Although the operator may take no direct part in the mooring operation, he should have a representative on board the drilling rig. The operator has a direct interest in seeing that the mooring system is well designed, well placed, and effective, inasmuch as he will pay for any down time experienced by the rig due to adverse weather. The second company involved is the drilling contractor, who owns the rig. The drilling contractor is represented by a drilling superintendent, a barge engineer, and/or a ship captain, in the case of a self-propelled vessel. The drilling superintendent has charge of the vessel during normal drilling operations, while the barge engineer is responsible for the stability and integrity of the vessel. Anchor handling boats and the masters and mates are supplied by the operator and usually constitute the third company. Finally, the anchor handling crews who establish the mooring system are a highly skilled, specialized fourth company. In foreign operations where language can be a problem, it may be desirable to have a representative of the anchor handling crew on board the drilling rig and vice versa.

MOVING ONTO LOCATION

A survey boat marks the well site and the anchor locations with marker buoys before the drilling rig arrives on location. In certain cases, radar enables the rig to set the anchor locations without marker buoys, but this is not usual practice.

For a local, or in-field move, the rig is usually towed onto location using one of the pendant lines as a tow line. Line 2 in figure 52 exemplifies this method. If the tow is a long, ocean tow under rough conditions, a heavier line is used.

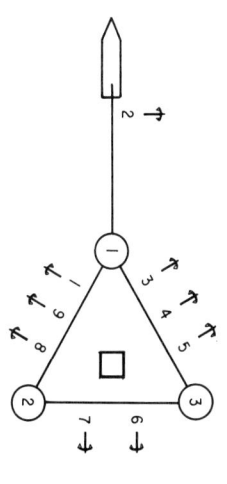

Figure 52. Attachment Locations for Pendant Lines to Main Deck

Figure 53 illustrates the typical sequence of running anchors for a SEDCO 135 rig. As the rig moves onto location, two stern anchors (Nos. 6 and 7) are dropped at the marker buoy, and mooring chain is paid out by the windlasses as the rig moves ahead. The windlasses are stopped when the rig has reached the location. The anchor handling boat then returns to the rig and runs out the bow anchor (No. 2). Remaining anchors are then run, with those in the upwind and upcurrent directions coming first.

LEGEND

⚓ Anchor
+ Buoy Position
} Marker Buoys

Setting Order
6 and 7
2, 1, 3, 5, 8, 9, 4

Retrieving Order
9, 4, 1, 3, 5, 8, 6, 7, 2
(Depending on Sea Conditions)

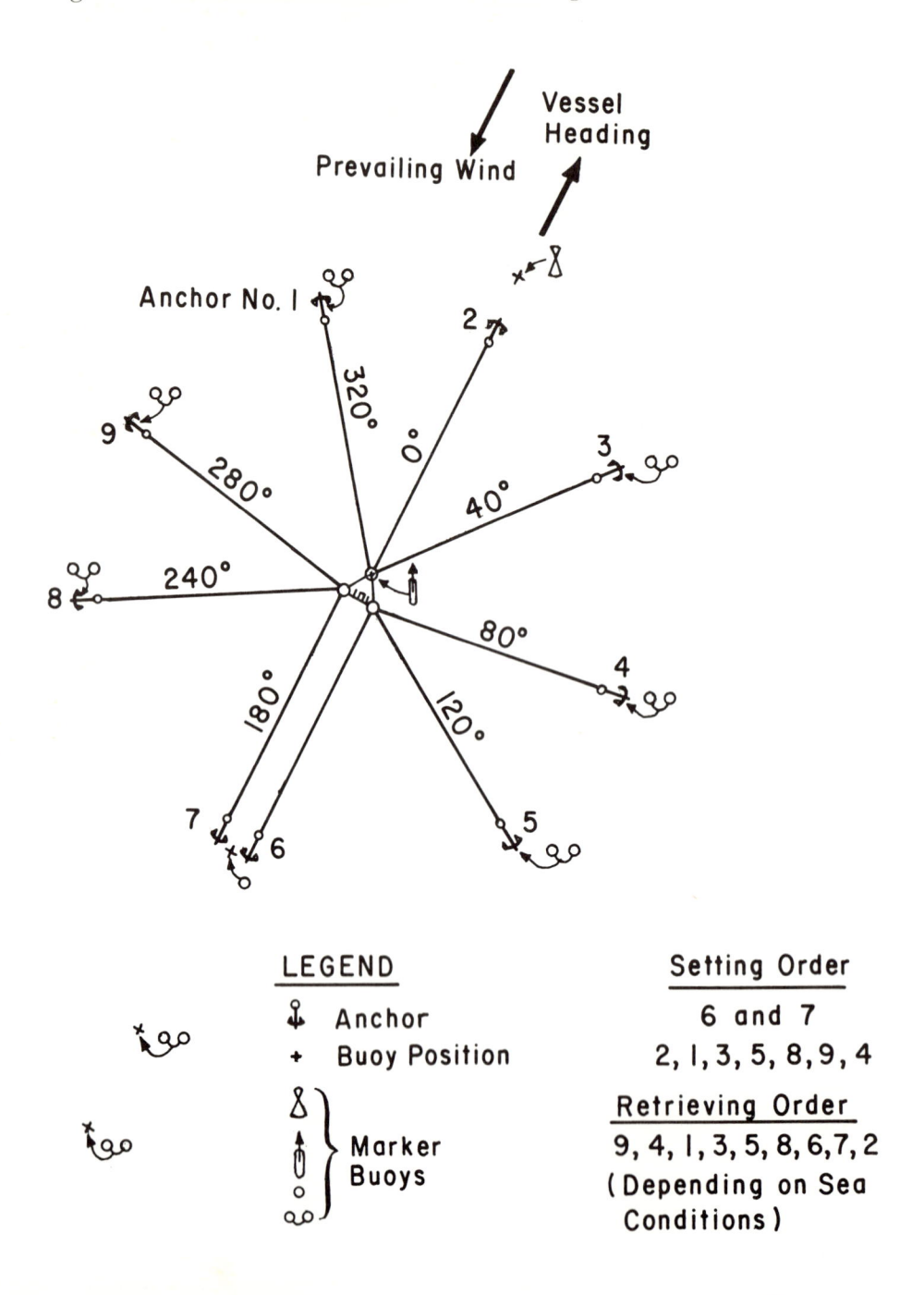

Figure 53. Mooring Pattern and Sequence of Setting and Retrieving Anchors

RUNNING OUT AN ANCHOR

To run out an anchor, the anchor handling boat initially comes to a location where the pendant line for the designated anchor can be passed from the rig to the boat. The anchor handling boat maneuvers into a position with its stern towards the rig and takes a heading towards the appropriate marker buoy. On some vessels, the rig cranes can be used to pass the pendant lines; on others, it is necessary to use heaving lines. After the pendant line has been passed over the tail roller of the anchor handling boat, it is attached to the deck winch and the slack is taken up. When the pendant line is securely on the deck winch, the boat advises the rig that it is ready to run the anchor. The anchor is then lowered, mooring line is paid out, and the boat pulls away.

Success of this operation and the maximum sea conditions under which it can be achieved are a function of the power and equipment of the anchor handling boat and the competence of its crew. The greatest danger is the possibility of collision between the boat and the rig. To help with maneuvering, the anchor handling boat should be equipped with bow thrusters or supplied with a tug for assistance. As it maneuvers, the anchor handling boat should not back over the pendant line and the anchor because this can foul the propellers. If a crane operator has control of the mooring line, he can swing the line away if necessary.

When the anchor handling boat has loaded the anchor and pulled away from the rig, the weight of the anchor and mooring line should be snubbed off with pelican hooks and not held by pulling the deck winch. The speed of the boat when carrying out the anchor is controlled by the rate of payout from the winch or windlass on the rig. If the mooring line is chain, a high payout rate could cause a pile-up on the bottom. The hazard of kinks and snags, along with increasing drag on the anchor handling boat must also be avoided. The anchor handling boat can decrease drag by keeping the anchor pulled up to the stern of the boat as close as is practical. Determination of the correct payout speed depends on the experience of the winch operator and his ability to control the tension in the chain. Increase in tension should be minimized to avoid unnecessary pull on the boat; however, sufficient tension should be maintained to prevent pile-ups on bottom. Tension indicators for chain tend to be inaccurate when it is run dynamically; therefore, many operators rely on the ampere-meter reading of the windlass electric motor, i.e., its torque, as a measure of chain tension. There is an incentive to run chain as fast as possible in order to build up momentum which will carry it to full length. The windlass should be monitored closely to make sure that the chain does not jump out of the wildcat at high speeds.

SETTING AN ANCHOR

When the anchor handling boat reaches the marker buoy, the anchor is lowered to the bottom and then lifted 30 feet. The anchor is held at this level while the rig starts heaving in (pulling) to eliminate any slack from the mooring line. When the line is taut, further heaving in on the line pulls the anchor handling boat away from the marker buoy. At this time, the anchor is lowered with the flukes down and set, as shown in figure 54. In soft bottom conditions, the flukes will trip and the anchor will set.

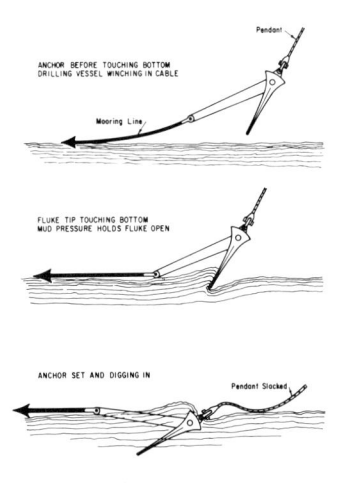

Figure 54. Sequence for Properly Setting Anchor from Anchor Handling Boat

PRELOADING AND PRETENSIONING ANCHOR LINES

Once all the anchors in a mooring pattern have been run and set, they are preloaded by tensioning the lines to a maximum of one-half breaking strength. This is above the design "working load" of a mooring line and normally it should exceed the holding power of the anchors. Full tension should be maintained for 10 to 15 minutes; however, experienced personnel know immediately if an anchor will hold. Readings on the tension indicator will slowly bleed back. Holding power of an anchor can be improved sometimes by allowing it to "soak" for awhile. The bottom soil tends to consolidate after being disturbed, and resistance improves.

After all the anchors have been preloaded and tested, the mooring lines are pretensioned to the static forces which they will carry on a regular, continuing basis. Pretension depends on the water depth and environmental conditions in the area. Usually, 50,000 lbs. pretension is the minimum allowable, increasing to over 200,000 lbs. in rough areas.

PIGGYBACKING ANCHORS

If an anchor will not hold when preloaded, "piggybacking," or installation of a second anchor on one mooring line, may be necessary, as shown in figure 55. The anchor handling boat attaches and sets the second anchor at the end of the mooring line. Actual installation procedures vary from one contractor to another. Many contractors simply detach the mooring buoy from the pendant line of the first anchor and reattach the pendant line to a second anchor. An additional pendant line is attached to the second anchor, and this line is used to lower and set the second anchor while keeping the line to the first anchor tightly stretched. Other contractors prefer to pull the first anchor on deck of the anchor handling boat, detach the pendant line, and reattach a heavier line — as much as three inches in diameter — for the second anchor. This line is used to relower and set the first anchor. The heavy line from the first anchor and a pendant line are attached to the second anchor, and it, too, is lowered and set as before.

MOORING SYSTEM ADJUSTMENTS IN ROUGH WEATHER

During storm conditions, tension in lines on the windward side of the vessel may increase significantly. When tensions which were higher than preload values occur consistently, reductions are achieved by slacking the leeward lines. Adjustments should then be made to equalize the tensions between the windward lines and the leeward lines as much as possible while remaining over the well. A measured approach based on experience is necessary, inasmuch as slacking the leeward lines permits the vessel to drift further from the well location.

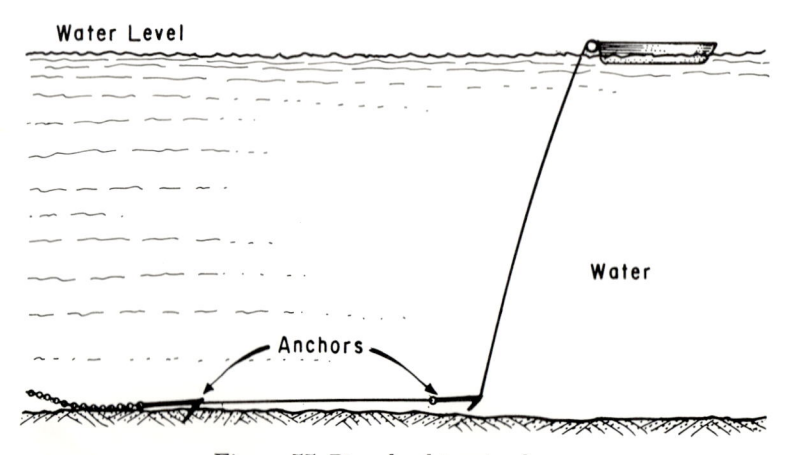

Figure 55. Piggybacking Anchors

If extreme weather conditions are expected, evacuation of the vessel may be necessary. All lines should be slacked off to the minimum level, approximately 50,000 lbs., and lines on the windward side should be run out close to the maximum length available.

PULLING AN ANCHOR

Before pulling an anchor, the rig slacks off all tension on the mooring line. The anchor handling boat approaches the mooring buoy, passes a line around the cross on top, and winches the buoy on board over the tail roller. The pendant wire is snubbed off with pelican hooks while the buoy is released and stowed on deck.

One of two procedures may then be followed to pull the anchor. In the first procedure, the pendant wire is made taut and the heave of the boat is used to work the anchor loose, as shown at left in figure 56. This procedure is extremely dangerous because it places severe shock loads on the pendant line and may cause the line to snap. If the line does snap, the resultant whiplash could cause severe injury to nearby personnel. The procedure shown at right (figure 56) is preferred, wherein the boat pulls away on the pendant line under power until the anchor comes loose. This second procedure places a steadier load on the pendant line, minimizing its possibility to snap.

Once the anchor is loose, it should be raised to the surface to see that it is free and clear. Caution while raising the anchor assures the flukes are away from the boat as it surfaces. If the flukes are toward the boat, it is sometimes possible to turn them away with propeller wash. When the anchor is free and clear, the mooring line is heaved in by the rig until the anchor can be hung on the anchor rack. In the meantime, the pendant line has been "laced" on the deck of the anchor handling boat and is passed to the rig, along with the mooring buoy.

All components of the mooring system should be inspected as they are recovered. Wire ropes should be checked for corrosion, kinks, and broken wires. Connecting links, thimbles, and shackles should be free of cracks. As chain passes through the windlass, it should be checked for missing studs and bent links.

MOVING OFF LOCATION

The order of retrieving anchors is normally the reverse of the setting order. The port and starboard anchors are pulled first, the exact order depending on sea conditions. Leeward anchors came next, then the two stern anchors. The bow anchor is pulled last. After pulling the bow anchor, the anchor handling boat retains the pendant line and begins the tow off location. Figures 57 to 63 show the sequence of retrieving and loading anchors on board an anchor handling boat.

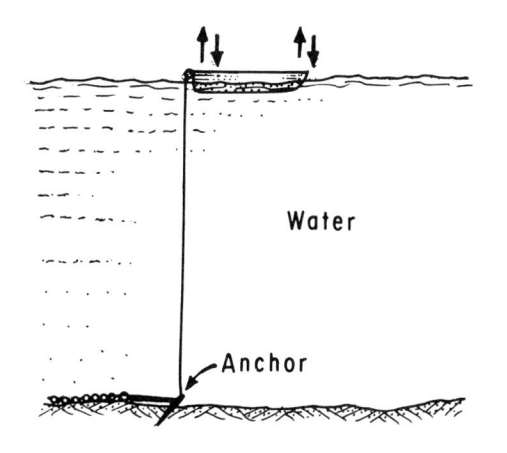

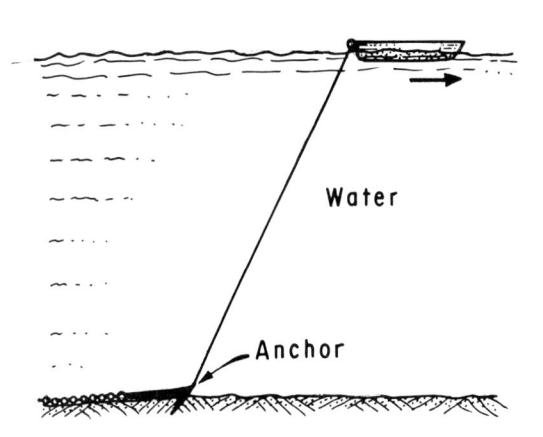

Figure 56. Heave of the Anchor Handling Boat

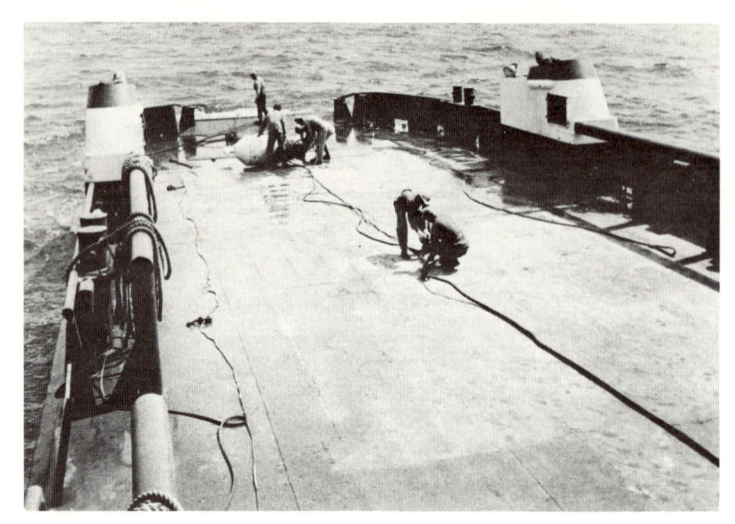

Figure 57. Buoy is Hauled on Deck

Figure 58. Buoy is Removed from Pendant Line

Figure 59. First Anchor to the Surface

Figure 60. First Anchor Pulled Up and Over the Rolling Tail Piece

Figure 61. The Winch that Does the Work to Pick Up the Anchors

Figure 62. Eight Anchors Retrieved and on Deck

Figure 63. Anchors Loaded and Ready for Rig Move

GLOSSARY

balling up: failure of an anchor to hold in soft bottom, when it pulls out with a large ball of mud

breaking strength: the load at which a chain parts

catenary curve: the shape assumed by a perfectly flexible line hanging under its own weight

crown chain: a length of chain between the crown of an anchor and the end of the pendant line

fairlead: a type of sheave which is used to change the direction of a moving line

lay: manner in which the individual wires and strands of a rope are wound together

leeward: the downwind direction

pelican hook: a device releasable under tension which engages chain links and wire rope fittings

piggybacking: installing tandem anchors on the same mooring line

preload: the test load imposed on a mooring line after it has been run

pretension: static load carried by a mooring line on a regular, continuing basis

proof load: a load slightly higher than the yield strength of the steel in a chain

scope: the ratio of total length of a mooring line to water depth

stud-link chain: a type of oval-link anchor chain with a bar, or stud, across the short dimension of the link to prevent tangling and deformation under load

tail roller: large roller across the stern of an anchor handling boat

whelps: sprocket teeth in a wildcat

wildcat: gear-like sheave of windlass for pulling anchor chain

windlass: propels chain from and to a chain locker where it is stored

windward: the upwind direction

winch: winds wire rope off and onto a storage drum, and provided with a brake to control the speed of unwinding the wire rope

REFERENCES

Acaster, S. M., "Continuing Wire Ropes for Anchoring Semisubmersibles," *Journal of Petroleum Technology*, June, 1973, pp. 659-662

Adams, Roy B., "Analysis of Spread Moorings by Dimensionless Function," Offshore Technology Conference Paper Number OTC 1077, 1969

American Petroleum Institute, "API Specifications for Mooring Chain," API Spec. 2F, latest edition

Beck, R. W., "Anchor Performance Tests," Offshore Technology Conference Paper Number 1537, 1972

Childers, M.A., "Mooring Systems for Hostile Waters," *Petroleum Engineer*, May, 1973, pp. 58-70 — "Mooring Systems," ETA Seminar, Inc., 1972

Collipp, B. G., "Analyzing Mooring Line Catenaries," *Petroleum Engineer*, 40,5, May, 1968

"History and Development of Anchor Chain," — "History and Development of the Marine Anchor," Baldt Chain Corporation

Howe, Richard J., "Development of Offshore Drilling and Production Technology," Proceedings of the ASME Underwater Technology Division Conference, Newport News, Virginia, April 30 - May 3, 1967

Korkut, M.D. and E. J. Hebert, "Some Notes on Static Anchor Chain Curve," Offshore Technology Conference Paper Number 1160, 1970

Stanton, P.N., D. R. Tidwell, and J. R. Lloyd, "Sloping Sea Floors," Offshore Technology Conference Paper Number OTC 1158, 1970

Wire Rope Engineering Handbook, American Steel and Wire Division, United States Steel, 1946

Wire Rope Handbook, Armco Steel Corporation, 1972

ACKNOWLEDGMENTS

The material presented in this manual was prepared by James E. Dailey of The University of Texas at Austin and represents an accumulation of experience and suggestions from a number of people who have devoted many years to the floating drilling industry. Among those whose contributions are gratefully acknowledged are Mr. Allan Bryant, Mr. Mark Childers, Mr. Dillard Hammett, Mr. Richard Knapp, Mr. Charles Linnenbank, and Mr. Bobby Lynch.